Principles of Food and Beverage Management

Second Edition

ManageFirst PROGRAM®

NATIONAL RESTAURANT ASSOCIATION®

PEARSON

Boston Columbus Indianapolis New York San Francisco Upper Saddle River
Amsterdam Cape Town Dubai London Madrid Milan Munich Paris Montréal Toronto
Delhi Mexico City São Paulo Sydney Hong Kong Seoul Singapore Taipei Tokyo

Pearson

Editorial Director: Vernon R. Anthony
Executive Acquisitions Editor: Alli Gentile
NRA Product Development: Randall Towns and Todd Schlender
Senior Managing Editor: JoEllen Gohr
Associate Managing Editor: Alexandrina B. Wolf
Senior Operations Supervisor: Pat Tonneman
Senior Operations Specialist: Deidra Skahill
Cover photo: Haken Dare/iStockPhoto
Cover design: Karen Steinberg, Element LLC
Director of Marketing: David Gesell
Senior Marketing Manager: Thomas Hayward
Marketing Coordinator: Les Roberts
Full-Service Project Management: Barbara Hawk and Kevin J. Gray, Element LLC
Text and Cover Printer/Binder: LSC Communications
Text Font: Minion Pro, Myriad Pro Semicondensed

Photography Credits

Front matter: i Haken Dare/iStockPhoto; vii (left) Suhendri Utet/Dreamstime; (right) Meryll/Dreamstime; viii (top) Mtr/Dreamstime; (bottom) Stratum/Dreamstime; ix (bottom left) Aprescindere/Dreamstime; xv (bottom left) Petar Neychev/Dreamstime; 34, 67, 190, 222, 306 Nikada/iStockPhoto

All other photographs owned or acquired by the National Restaurant Association Educational Foundation, NRAEF

---

13 2022

PEARSON

NATIONAL RESTAURANT ASSOCIATION®

ISBN-10: 0-13-272549-5
ISBN-13: 978-0-13-272549-1

ISBN-10: 0-13-274205-5
ISBN-13: 978-0-13-274205-4

Contents in Brief

Contents

About the National Restaurant Association and the National Restaurant Association Educational Foundation

Founded in 1919, the National Restaurant Association (NRA) is the leading business association for the restaurant and foodservice industry, which comprises 960,000 restaurant and foodservice outlets and a workforce of nearly 13 million employees. We represent the industry in Washington, DC, and advocate on its behalf. We operate the industry's largest trade show (NRA Show, restaurant.org/show); leading food safety training and certification program (ServSafe, servsafe.com); unique career-building high school program (the NRAEF's *ProStart*, prostart.restaurant.org); as well as the *Kids LiveWell* program (restaurant.org/kidslivewell) promoting healthful kids' menu options. For more information, visit www.restaurant.org and find us on Twitter *@WeRRestaurants, Facebook,* and *YouTube.*

With the first job experience of one in four U.S. adults occurring in a restaurant or foodservice operation, the industry is uniquely attractive among American industries for entry-level jobs, personal development and growth, employee and manager career paths, and ownership and wealth creation. That is why the National Restaurant Association Educational Foundation (nraef.org), the philanthropic foundation of the NRA, furthers the education of tomorrow's restaurant and foodservice industry professionals and plays a key role in promoting job and career opportunities in the industry by allocating millions of dollars a year toward industry scholarships and educational programs. The NRA works to ensure the most qualified and passionate people enter the industry so that we can better meet the needs of our members and the patrons and clients they serve.

What Is the ManageFirst Program?

The ManageFirst Program is a management training certificate program that exemplifies our commitment to developing materials by the industry, for the industry. The program's

EXAM TOPICS

ManageFirst Core Credential Topics
Hospitality and Restaurant Management
Controlling Foodservice Costs
Hospitality Human Resources Management and Supervision
ServSafe® Food Safety

ManageFirst Foundation Topics
Customer Service
Principles of Food and Beverage Management
Purchasing
Hospitality Accounting
Bar and Beverage Management
Nutrition
Hospitality and Restaurant Marketing
ServSafe Alcohol® Responsible Alcohol Service

most powerful strength is that it is based on a set of competencies defined by the restaurant and foodservice industry as critical for success. The program teaches the skills truly valued by industry professionals.

ManageFirst Program Components

The ManageFirst Program includes a set of books, exams, instructor resources, certificates, a new credential, and support activities and services. By participating in the program, you are demonstrating your commitment to becoming a highly qualified professional either preparing to begin or to advance your career in the restaurant, hospitality, and foodservice industry.

These books cover the range of topics listed in the chart above. You will find the essential content for the topic as defined by industry, as well as learning activities, assessments, case studies, suggested field projects, professional profiles, and testimonials. The exam can be administered either online or in a paper-and-pencil format (see inside front cover for a listing of ISBNs), and it will be proctored. Upon successfully passing the exam, you will be furnished with a customized certificate by the National Restaurant Association. The certificate is a lasting recognition of your accomplishment and a signal to the industry that you have mastered the competencies covered within the particular topic.

To earn this credential, you will be required to pass four core exams and one foundation exam (to be chosen from the remaining program topics) and to document your work experience in the restaurant and foodservice industry. Earning the ManageFirst credential is a significant accomplishment.

We applaud you as you either begin or advance your career in the restaurant, hospitality, and foodservice industry. Visit www.nraef.org to learn about additional career-building resources offered by the NRAEF, including scholarships for college students enrolled in relevant industry programs.

MANAGEFIRST PROGRAM ORDERING INFORMATION

Review copies or support materials

FACULTY FIELD SERVICES
Tel: 800.526.0485

Domestic orders and inquiries

PEARSON CUSTOMER SERVICE
Tel: 800.922.0579
http://www.pearsonhighered.com/

International orders and inquiries

U.S. EXPORT SALES OFFICE
Pearson Education International Customer Service Group
200 Old Tappan Road
Old Tappan, NJ 07675 USA
Tel: 201.767.5021
Fax: 201.767.5625

For corporate, government, and special sales (consultants, corporations, training centers, VARs, and corporate resellers) orders and inquiries

PEARSON CORPORATE SALES
Tel: 317.428.3411
Fax: 317.428.3343
Email: managefirst@prenhall.com

For additional information regarding other Pearson publications, instructor and student support materials, locating your sales representative, and much more, please visit *www.pearsonhighered.com/managefirst*.

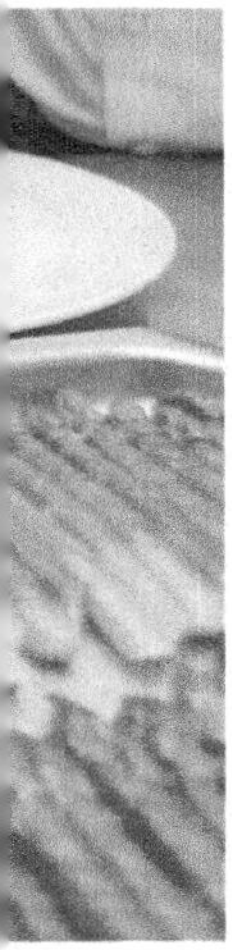

Acknowledgements

The National Restaurant Association is grateful for the significant contributions made to this book by the following individuals.

Mike Amos
Perkins & Marie Callender's Inc.

Steve Belt
Monical's Pizza

Heather Kane Haberer
Carrols Restaurant Group

Erika Hoover
Monical's Pizza Corp.

Jared Kulka
Red Robin Gourmet Burgers

Tony C. Merritt
Carrols Restaurant Group

H. George Neil
Buffalo Wild Wings

Marci Noguiera
Sodexo—Education Division

Ryan Nowicki
Dave & Busters

Penny Ann Lord Prichard
Wake Tech/NC Community College

Michael Santos
Micatrotto Restaurant Group

Heather Thitoff
Cameron Mitchell Restaurants

Features of the ManageFirst books

We have designed the ManageFirst books to enhance your ability to learn and retain important information that is critical to this restaurant and foodservice industry function. Here are the key features you will find within this book.

BEGINNING EACH BOOK

Real Manager

This is your opportunity to meet a professional who is currently working in the field associated with the book's topic. This person's story will help you gain insight into the responsibilities related to his or her position, as well as the training and educational history linked to it. You will also see the daily and cumulative impact this position has on an operation, and receive advice from a person who has successfully met the challenges of being a manager.

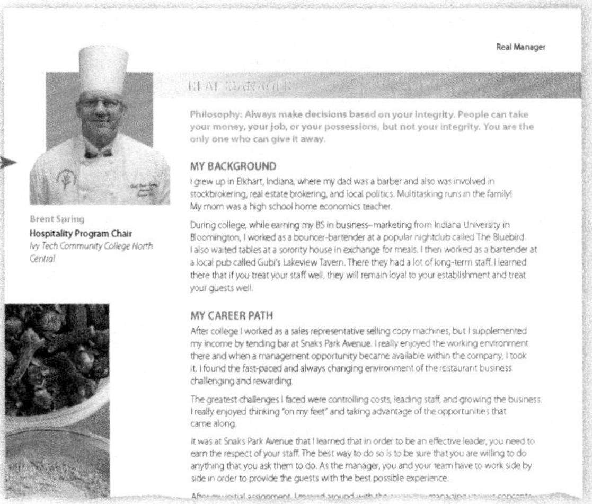

Real Manager

REAL MANAGER

Philosophy: Always make decisions based on your integrity. People can take your money, your job, or your possessions, but not your integrity. You are the only one who can give it away.

Brent Spring
Hospitality Program Chair
Ivy Tech Community College North Central

MY BACKGROUND

I grew up in Elkhart, Indiana, where my dad was a barber and also was involved in stockbrokering, real estate brokering, and local politics. Multitasking runs in the family! My mom was a high school home economics teacher.

During college, while earning my BS in business–marketing from Indiana University in Bloomington, I worked as a bouncer-bartender at a popular nightclub called The Bluebird. I also waited tables at a sorority house in exchange for meals. I then worked as a bartender at a local pub called Gubi's Lakeview Tavern. There they had a lot of long-term staff. I learned there that if you treat your staff well, they will remain loyal to your establishment and treat your guests well.

MY CAREER PATH

After college I worked as a sales representative selling copy machines, but I supplemented my income by tending bar at Snaks Park Avenue. I really enjoyed the working environment there and when a management opportunity became available within the company, I took it. I found the fast-paced and always changing environment of the restaurant business challenging and rewarding.

The greatest challenges I faced were controlling costs, leading staff, and growing the business. I really enjoyed thinking "on my feet" and taking advantage of the opportunities that came along.

It was at Snaks Park Avenue that I learned that in order to be an effective leader, you need to earn the respect of your staff. The best way to do so is to be sure that you are willing to do anything that you ask them to do. As the manager, you and your team have to work side by side in order to provide the guests with the best possible experience.

BEGINNING EACH CHAPTER

Inside This Chapter

Chapter content is organized under these major headings.

Learning Objectives

Learning objectives identify what you should be able to do after completing each chapter. These objectives are linked to the required tasks a manager must be able to perform in relation to the function discussed in the book.

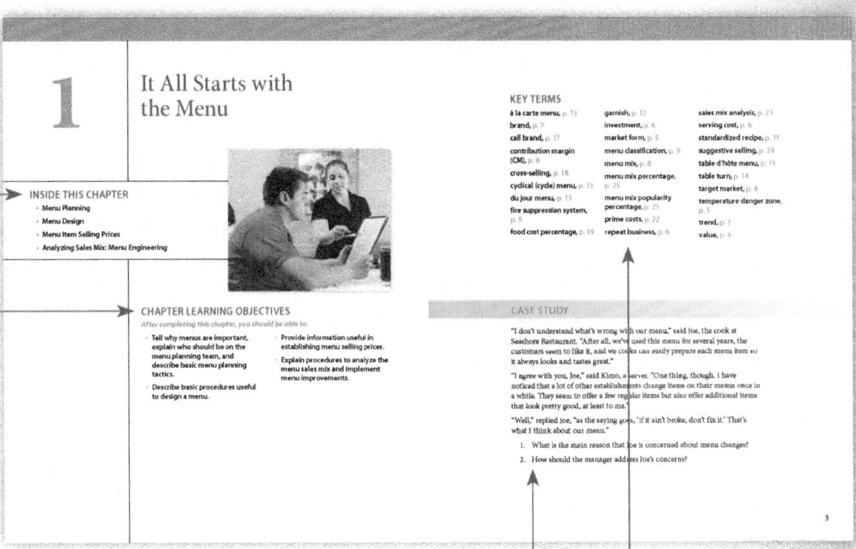

1 It All Starts with the Menu

INSIDE THIS CHAPTER
- Menu Planning
- Menu Design
- Menu Item Selling Prices
- Analyzing Sales Mix: Menu Engineering

CHAPTER LEARNING OBJECTIVES

After completing this chapter, you should be able to:
- Tell why menus are important, explain who should be on the menu planning team, and describe basic menu planning tactics.
- Describe basic procedures useful to design a menu.
- Provide information useful in establishing menu selling prices.
- Explain procedures to analyze the menu sales mix and implement menu improvements.

KEY TERMS

à la carte menu, p. 13
brand, p. 7
call brand, p. 17
contribution margin (CM), p. 8
cross-selling, p. 18
cyclical (cycle) menu, p. 13
du jour menu, p. 13
fire suppression system, p. 5
food cost percentage, p. 19
garnish, p. 12
investment, p. 6
market form, p. 5
menu classification, p. 9
menu mix, p. 8
menu mix percentage, p. 25
menu mix popularity percentage, p. 25
prime costs, p. 22
repeat business, p. 6
sales mix analysis, p. 23
serving cost, p. 6
standardized recipe, p. 11
suggestive selling, p. 29
table d'hôte menu, p. 13
table turn, p. 14
target market, p. 4
temperature danger zone, p. 5
trend, p. 7
value, p. 4

CASE STUDY

"I don't understand what's wrong with our menu," said Joe, the cook at Seashore Restaurant. "After all, we've used this menu for several years, the customers seem to like it, and we cooks can easily prepare each menu item so it always looks and tastes great."

"I agree with you, Joe," said Kimo, a server. "One thing, though. I have noticed that a lot of other establishments change items on their menus once in a while. They seem to offer a few regular items but also offer additional items that look pretty good, at least to me."

"Well," replied Joe, "as the saying goes, 'If it ain't broke, don't fix it.' That's what I think about our menu."

1. What is the main reason that Joe is concerned about menu changes?
2. How should the manager address Joe's concerns?

3

Case Study

Each chapter begins with a brief story about the kind of situations that a manager may encounter in the course of his or her work. The story is followed by one or two questions to prompt student discussions about the topics contained within the chapter.

Key Terms

These terms are important for thorough understanding of the chapter's content. They are highlighted throughout the chapter, where they are explicitly defined or their meaning is made clear within the paragraphs in which they appear.

THROUGHOUT EACH CHAPTER

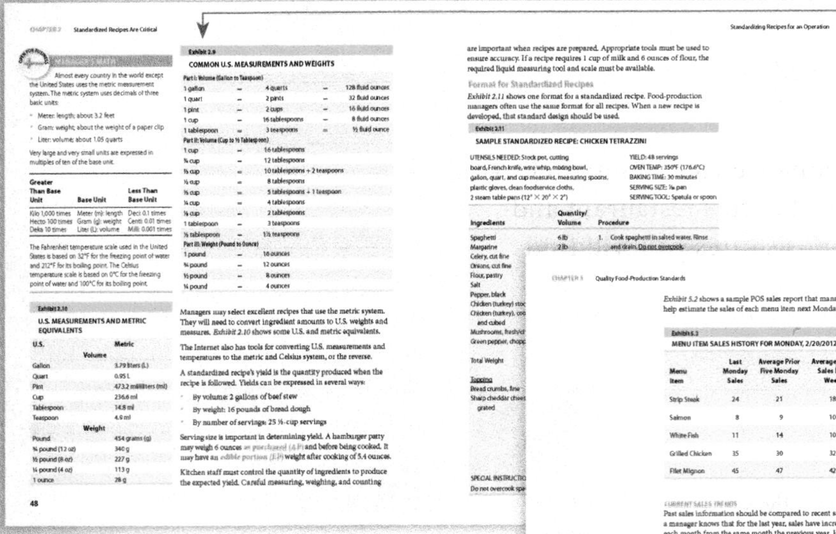

Exhibits

Exhibits are placed throughout each chapter to visually reinforce the key concepts presented in the text. Types of exhibits include charts, tables, photographs, and illustrations.

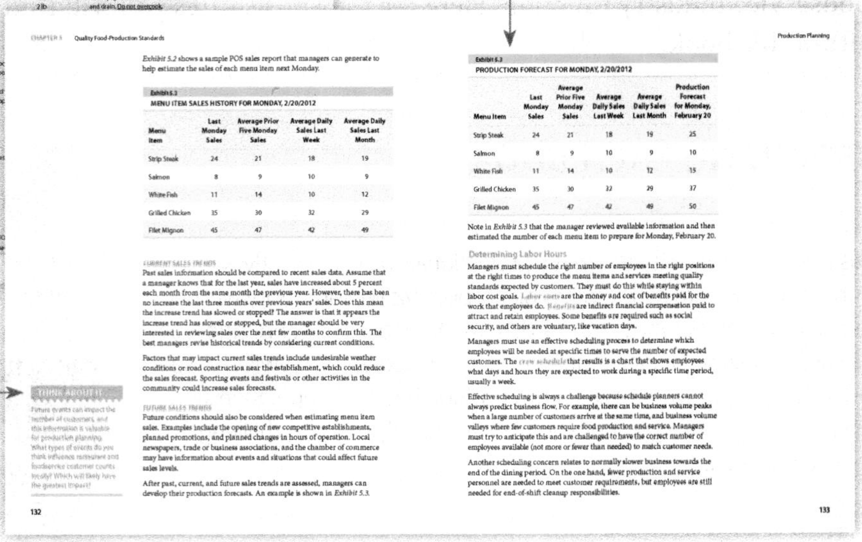

Think About It . . .

These thought-provoking sidebars reveal supportive information about the section they appear beside.

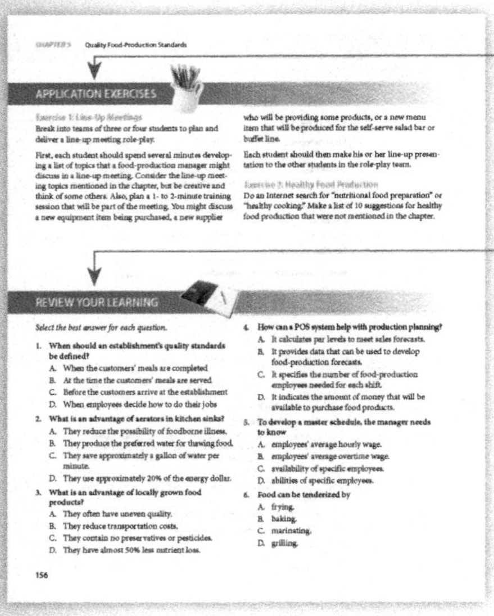

AT THE END OF EACH CHAPTER

Application Exercises and Review Your Learning

These multiple-choice or open- or close-ended questions or problems are designed to test your knowledge of the concepts presented in the chapter. These questions have been aligned with the objectives and should provide you with an opportunity to practice or apply the content that supports these objectives. If you have difficulty answering the Review Your Learning questions, you should review the content further.

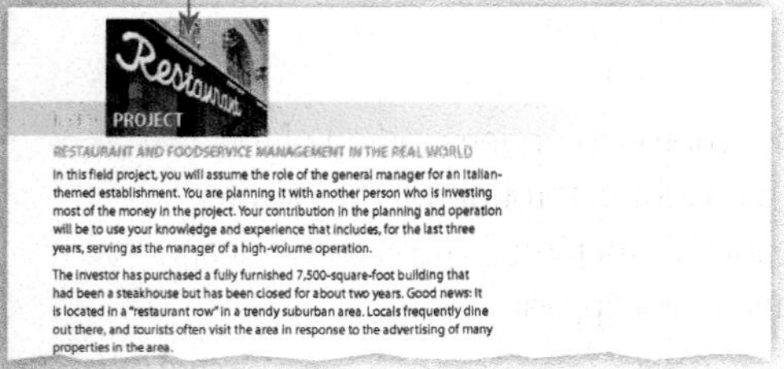

AT THE END OF THE BOOK

Field Project

This real-world project gives you the valuable opportunity to apply many of the concepts you will learn in a competency guide. You will interact with industry practitioners, enhance your knowledge, and research, apply, analyze, evaluate, and report on your findings. It will provide you with an in-depth "reality check" of the policies and practices of this management function.

REAL MANAGER

Brent Spring
Hospitality Program Chair
Ivy Tech Community College North Central

Philosophy: Always make decisions based on your integrity. People can take your money, your job, or your possessions, but not your integrity. You are the only one who can give it away.

MY BACKGROUND

I grew up in Elkhart, Indiana, where my dad was a barber and also was involved in stockbrokering, real estate brokering, and local politics. Multitasking runs in the family! My mom was a high school home economics teacher.

During college, while earning my BS in business–marketing from Indiana University in Bloomington, I worked as a bouncer-bartender at a popular nightclub called The Bluebird. I also waited tables at a sorority house in exchange for meals. I then worked as a bartender at a local pub called Gubi's Lakeview Tavern. There they had a lot of long-term staff. I learned there that if you treat your staff well, they will remain loyal to your establishment and treat your guests well.

MY CAREER PATH

After college I worked as a sales representative selling copy machines, but I supplemented my income by tending bar at Snaks Park Avenue. I really enjoyed the working environment there and when a management opportunity became available within the company, I took it. I found the fast-paced and always changing environment of the restaurant business challenging and rewarding.

The greatest challenges I faced were controlling costs, leading staff, and growing the business. I really enjoyed thinking "on my feet" and taking advantage of the opportunities that came along.

It was at Snaks Park Avenue that I learned that in order to be an effective leader, you need to earn the respect of your staff. The best way to do so is to be sure that you are willing to do anything that you ask them to do. As the manager, you and your team have to work side by side in order to provide the guests with the best possible experience.

After my initial assignment, I moved around with the company, managing various concepts. I became a turnaround specialist, and the company would move me from location to location to get restaurants back on track. At one of these locations, we had banquet rooms and off-site catering. I was in charge of managing the off-site events in addition to restaurant operations. At other locations , I was the bar manager, kitchen manager, dining-room manager, and general manager, as needed.

I then went to a hotel operation as director of restaurant and lounges for a property that had fine-dining, family dining, poolside club, comedy club, and nightclub operations. This position broadened my exposure to different hospitality environments.

At this point, my former company offered me the opportunity to become a regional director of operations. I oversaw the remodeling and concept change of several steakhouse restaurants as well as a sports bar and breakfast, lunch, and dinner family restaurant.

The position involved working on menu and recipe development, standardizing recipes and purchasing procedures, front of the house and kitchen operations procedure manual development, and profit and loss analysis. When the company sold the division I was working for, I became a district manager for the remaining division of 24-hour family dining restaurants. It was at this time that I also became an adjunct instructor in hospitality administration, teaching purchasing and cost control.

While I was a district manager, an opportunity to open my own banquet and catering business was presented to me. I decided to do it, since I wanted to travel less. I have been the president and executive chef of Feast Inc. for 10 years. When the economy took a downturn, my landlord decided not to renew my lease for the facility. This unfortunate turn of events provided an opportunity to change my career focus a bit.

At about the same time, the position of hospitality administration program chair became available at the community college at which I was teaching. I interviewed and was hired three years ago. I also have been involved with the American Culinary Federation in various leadership roles. I have held the offices of treasurer, vice president, and scholarship chair, and was voted Chef of the Year for the South Bend chapter in 2006 and 2011.

Something I always think about: **I've learned that—despite all the hard work you do—you work to live, you don't live to work. Your life needs to have balance in order to be objective and for you to be your best.**

WHAT DOES THE FUTURE HOLD?

I believe that opportunities in food and beverage management will continue to grow. As people lead busier lives, they will dine out and have food catered in more frequently. Those of you who take the time to learn the analysis process and make informed, data-driven decisions will have successful careers. Remember that decisions need to be made based on accurate data. If a decision is made based on feelings or opinions, it is less likely to be the right decision.

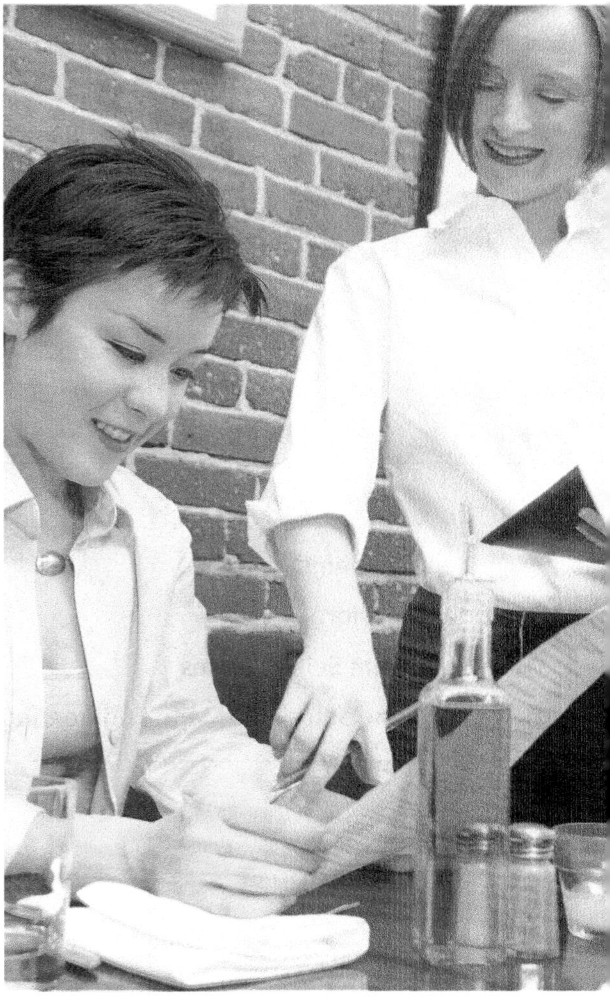

MY ADVICE TO YOU

You must be ready to work smart and work hard. Identify problems and present solutions to those problems. Success is dependent on your ability to come up with solutions to problems, then implement and execute the solutions.

Remember: **To be successful, you need to be able to not only recognize an opportunity but also be willing to act to seize that opportunity.**

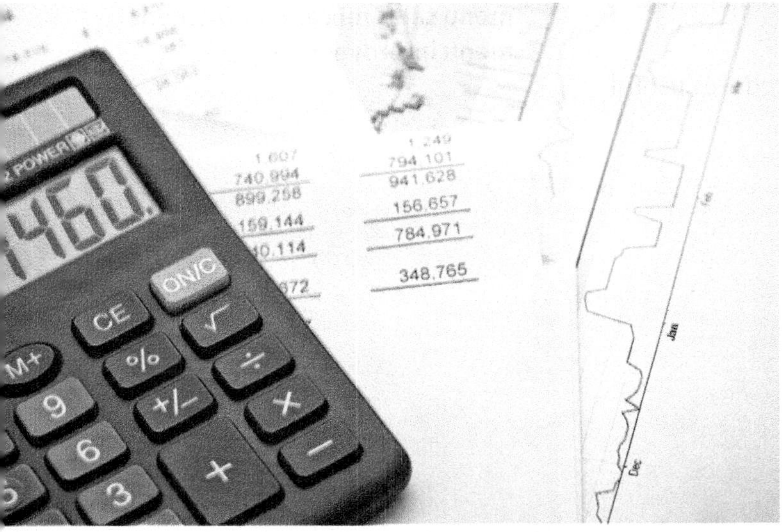

1 It All Starts with the Menu

INSIDE THIS CHAPTER

- Menu Planning
- Menu Design
- Menu Item Selling Prices
- Analyzing Sales Mix: Menu Engineering

CHAPTER LEARNING OBJECTIVES

After completing this chapter, you should be able to:

- Tell why menus are important, explain who should be on the menu planning team, and describe basic menu planning tactics.
- Describe basic procedures useful to design a menu.
- Provide information useful in establishing menu selling prices.
- Explain procedures to analyze the menu sales mix and implement menu improvements.

KEY TERMS

- **à la carte menu,** p. 13
- **brand,** p. 7
- **call brand,** p. 17
- **contribution margin (CM),** p. 8
- **cross-selling,** p. 18
- **cyclical (cycle) menu,** p. 13
- **du jour menu,** p. 13
- **fire suppression system,** p. 5
- **food cost percentage,** p. 19
- **garnish,** p. 12
- **investment,** p. 6
- **market form,** p. 5
- **menu classification,** p. 9
- **menu mix,** p. 8
- **menu mix percentage,** p. 25
- **menu mix popularity percentage,** p. 25
- **prime costs,** p. 22
- **repeat business,** p. 6
- **sales mix analysis,** p. 23
- **serving cost,** p. 6
- **standardized recipe,** p. 11
- **suggestive selling,** p. 29
- **table d'hôte menu,** p. 13
- **table turn,** p. 14
- **target market,** p. 4
- **temperature danger zone,** p. 5
- **trend,** p. 7
- **value,** p. 4

CASE STUDY

"I don't understand what's wrong with our menu," said Joe, the cook at Seashore Restaurant. "After all, we've used this menu for several years, the customers seem to like it, and we cooks can easily prepare each menu item so it always looks and tastes great."

"I agree with you, Joe," said Kimo, a server. "One thing, though. I have noticed that a lot of other establishments change items on their menus once in a while. They seem to offer a few regular items but also offer additional items that look pretty good, at least to me."

"Well," replied Joe, "as the saying goes, 'if it ain't broke, don't fix it.' That's what I think about our menu."

1. What is the main reason that Joe is concerned about menu changes?
2. How should the manager address Joe's concerns?

MENU PLANNING

The menu is an important tool that informs guests about the products they can purchase. Planning a menu involves much more than just deciding what food and beverage products should be offered. Menus are important because they drive the establishment's marketing and operating plans, which are critical to financial success. An effective menu planning team must be in place, and specific procedures are needed to plan the best possible menu.

Importance of Menu Planning

Why does the success of a restaurant or foodservice operation start with the menu? The menu attracts customers, impacts profits, and influences almost everything done during daily operations.

MENUS ATTRACT CUSTOMERS

The operation's customers are an important first concern when the menu is planned. What do they want? Why do they visit the establishment? How do they evaluate the food and the experience they receive? Menu planners must understand their target market to plan a menu customers will enjoy. The **target market** is a group of people with similar characteristics and similar demands of the marketplace. Ideally, a target market consists of all people with the desire and ability to purchase an operation's products and services.

Successful operations offer menus enjoyed by their customers. Customers will not return if they do not like the menu. When regular customers do return, they expect consistency in menu items, pricing, and portion size. This is why menu planners must determine and offer what their customers want.

Exhibit 1.1

Customers visit restaurant and foodservice operations for many reasons. However, they always want **value** (the relationship between selling price and quality) when they make dining-out decisions. The customers' views about value are influenced by the food and beverage products offered and by the service and atmosphere. In other words, customers judge their dining experience, and that is largely driven by the menu (*Exhibit 1.1*).

MENUS IMPACT FINANCIAL SUCCESS

All types of operations must control costs. Commercial restaurant and foodservice operations control costs in order to generate a profit from the sale of food and beverage products. Noncommercial operations have financial goals such as to generate a surplus (profit), to break even, or not to exceed a predetermined loss. A menu that offers products targeted to those most likely to visit the property and priced to provide value is critical to the financial success of every operation. Since they are so critically important, significant attention is required for menu planning and implementation.

MENUS IMPACT DAILY OPERATIONS

The menu directly impacts the operation's resources including employees, food products, time, and money. As a simple example, consider what is involved in the decision to offer French fries on a menu.

Frozen fries or fresh whole potatoes must be purchased, as must the frying oil. Freezer, refrigerator, or dry storage space will be needed depending on the product's **market form**: the way a food product is purchased such as frozen or fresh. Sanitation concerns become important particularly when products are produced in large quantities and held for later service; can this be done while keeping the product outside the temperature danger zone? The **temperature danger zone** is the temperature range at which most microorganisms grow best: 41°F to 135°F (5°C to 57°C).

Space in the kitchen, probably already limited, will be needed for the deep-fryer, which must be located under the proper ventilation and fire suppression systems (*Exhibit 1.2*). A **fire suppression system** contains chemicals that are automatically sprayed on equipment surfaces below the system if a fire begins. The amount of money needed to purchase the equipment, venting, and fire suppression system can be significant. Also, the proper utilities, such as gas and electric, must be available for the storage and deep-frying equipment.

Little labor is needed to open a bag of frozen French fries and put them in the deep-fryer. However, what if customers want a product made from whole potatoes? Then production personnel will need adequate skills and time to prepare the item.

This example shows that a menu planner's "simple" idea to add French fries to the menu can have a serious impact on many aspects of the operation.

Exhibit 1.2

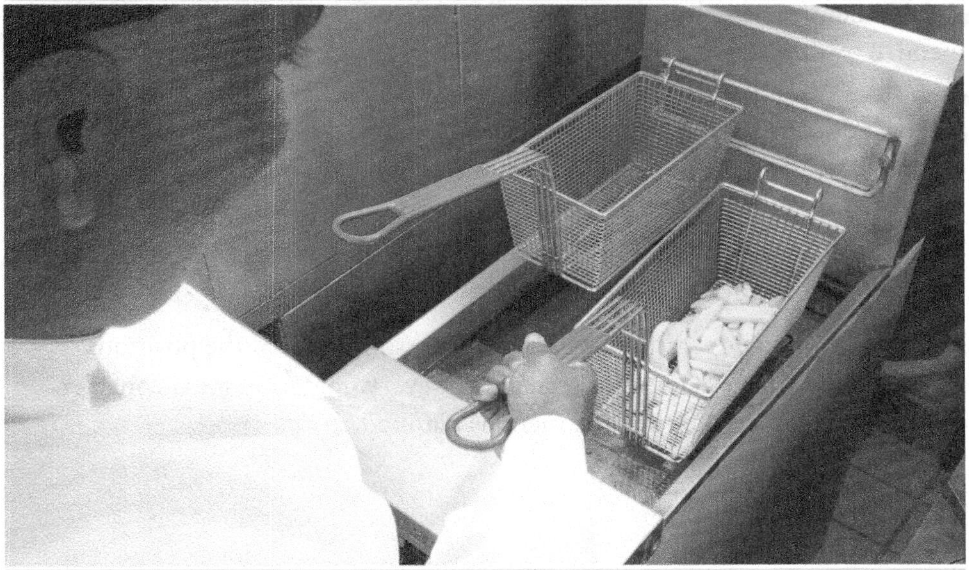

Manager's Memo

Almost every aspect of daily operations is affected by the menu items served. For example, the menu determines the recipes needed. They, in turn, drive the ingredients to be purchased and stored. Menu items must be prepared according to required quality standards. The kitchen's peak production capabilities must also be considered. What if a specific menu item becomes very popular, and the kitchen's capacity is exceeded?

The proper equipment must be available to produce and serve all menu items. Kitchen layout concerns are also important; kitchens that are too large because of equipment needed to deliver the menu can reduce productivity.

The dining areas should fit the type of menu items served. For example, Italian food would be out of place in a Mexican-themed establishment.

Kitchen employees must prepare menu items to desired quality levels. Likewise, service staff must serve the items in a manner that meets performance standards.

The Menu Planning Team

Exhibit 1.3 shows the positions that might be part of a menu planning team. It also indicates each position's special concerns as the menu is planned.

Exhibit 1.3

THE MENU PLANNING TEAM

Position	Menu Planning Concerns
Owner-manager	Will the menu help meet financial goals?
Kitchen manager	Will customers enjoy the items, and can items consistently be produced to the necessary quality and quantity standards?
Dining-room manager	Will the menu keep customers returning?
Purchasing agent	Are ingredients needed for menu items available at a reasonable price?
Accountant	Will financial goals be met? Some accountants help determine **serving cost** (the cost to produce one serving of a menu item prepared according to a standardized recipe) and selling price.

REAL MANAGER

THE MENU MATTERS!

After we had established ourselves as a casual restaurant, the executives of the company I worked for wanted to make our steakhouse concept more upscale. They redesigned the menu to give it a more upscale look. Then I was directed to implement the menu across all restaurants without first testing guests' response. The menu was not well received. As I worked at the locations I supervised, I found guests leaving the restaurants after looking at the menu. Our most popular items were not well placed and the most expensive items led off the menu. By observing guests' reactions and listening to employee feedback, we made some immediate adjustments to address these issues.

In reviewing *Exhibit 1.3,* remember that the owner-manager is concerned about the financial success of the operation. He or she wants to make a financial return that considers the risk of losing the **investment**: the amount of money an owner has used to start and operate the business.

The kitchen manager's team must be able to produce the items required by the menu. The dining-room manager wants to ensure that the menu helps build **repeat business**: revenue from customers who return because they enjoyed their experience during previous visits. He or she will know the popularity of existing menu items and have suggestions about other items that might be popular. Meanwhile, the purchasing agent should be able to advise about product availability, supplier sources, and price trends, and the accountant will have important financial concerns.

Only high-volume operations will employ persons in all of the positions shown in *Exhibit 1.3.* However, someone in every operation must consider each of these concerns when the menu is planned.

Menu Planning Steps

There are several steps that an operation must take in order to plan a successful menu. *Exhibit 1.4* reviews the planning steps that will be discussed in this section.

Exhibit 1.4

MENU PLANNING PROCEDURES

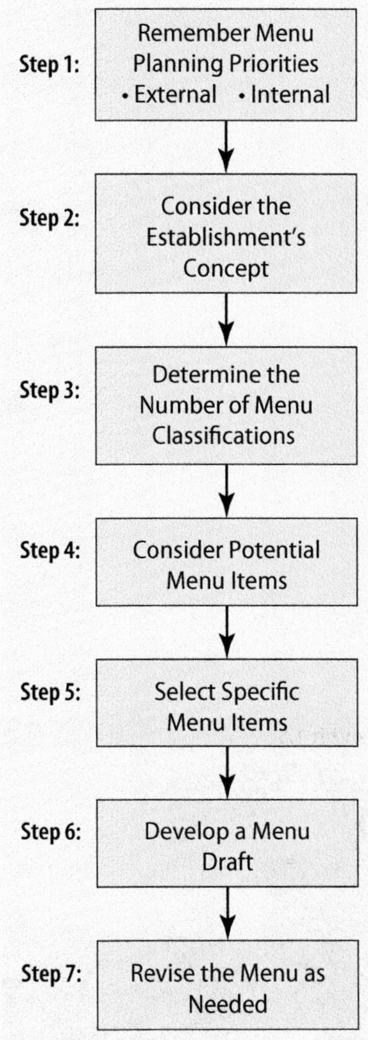

Step 1: Remember Menu Planning Priorities

Menu planners must consider external priorities that focus on why customers visit the property and internal priorities that consider the operation's ability to provide what customers want. Several external factors influence menu planning:

- **Target market:** Recall that a target market is the group of persons who might visit the property. Factors to consider include how often this group dines out, the price they are willing to pay, the environment and style of service they want, and the types of food they enjoy.
- **Competition:** An establishment's competitors are other operations that want to serve the same target market. Once competitors are identified, it is important to research their menu items and selling prices. An establishment might provide similar offerings but do so in a way that improves the item or provides a better dining experience. Possibilities include offering signature items that customers associate with a specific operation and setting higher standards for preparation, presentation, and service. Pricing is also important. Customers belong to a specific target market partly due to the price they are willing to pay for the perceived value they receive. To be competitive, prices on a planned menu should be similar to those of the competition.
- **Consumer trends:** Consumer trends affect the market for which a menu is planned. A **trend** is a gradual change in customers' food preferences, such as healthy eating, that is likely to continue for a significant time. A trend grows and may level off, but has high sales for a long period.
- **Brand:** A **brand** relates to how the establishment's products and services are different from its competitors'. Its purpose is to create an image in customers' minds that signifies a specific, desired perception of value for the price paid. A strong, well-defined brand is important to generate customer loyalty and visibility in the community. The physical brand may include a name, logo, symbol, or design. A brand's image may be reinforced by factors including exterior and interior decor, type of service, level of quality, type of menu items offered, menu price categories, and product consistency.

The menu affects almost all aspects of the establishment's internal operation. As the French fry example earlier in the chapter demonstrates, menu planners must consider the impact of menu items on a variety of internal areas:

- **Menu mix:** A new menu item might impact the **menu mix**: the frequency with which a menu item is ordered compared to other menu items. A change in menu mix could influence pre-preparation of food items as well as cooking.
- **Profit:** A new menu item might impact the operation's profitability. For example, its contribution margin might be lower than that of other items. The **contribution margin (CM)** is the difference between an item's revenue (selling price) and food cost. It is made up of profit and all expenses except food. If the contribution margin is too low, profit is reduced. Or, the food item might attract customers away from the high-profit items. In both cases, the net profit would be reduced.
- **Kitchen:** The capacity of kitchen preparation areas must be considered. If kitchen employees are already overloaded at the sauté station, for example, adding a new sauté item could have a negative impact on production.
- **Ingredients:** The availability of ingredients for the new menu item must be considered. A supplier source will be needed, and storage space will be required. If the item requires totally new ingredients, the size and expense of inventory will increase.
- **Equipment:** A major change in menu offerings may require the purchase and installation of a new piece of preparation equipment. In addition to purchase costs, it is often difficult to find space in the production line. Many pieces of cooking equipment need electricity, gas, water, and drainage connections, and possibly a ventilation system (*Exhibit 1.5*).
- **Production staff:** A menu change may require retraining of production employees. If the item is difficult or time consuming to produce, cooks may become frustrated. Since labor costs must always be considered, labor hours and skill levels for production are important considerations.
- **Service staff:** Service may be affected if the new menu item is served in special dishware, if it needs extra service ware, or if production slows service flow. Retraining of service employees may be needed so they can describe, and answer questions about, the new menu item.

Exhibit 1.5

Step 2: Consider the Establishment's Concept

What are customers' expectations about the menu based on their age, occupation, family status, ethnicity, and other factors addressed in the property's brand? New items should fit within these brand expectations.

All menu items must also meet the operation's quality requirements. They should reflect what customers want and are willing to pay for. Quality factors include each menu item's flavor, texture, form, shape, nutritional content, visual and aromatic (smell) appeal, and temperature. It should be noted that customers' views about quality can change over time, which may in turn require changes to the menu.

THINK ABOUT IT . . .

Many menu planners believe that if they take care of customers, financial goals will take care of themselves. Others focus on financials first. What would be the most important considerations to you as a menu planner?

Step 3: Determine the Number of Menu Classifications

Exhibit 1.4 indicates that the third step in menu planning requires planners to consider **menu classifications**: similar groups of items such as entrées and soups and salads.

Menu classifications typically include the following categories:

- **Entrées (main course):** These menu items are typically planned before others. Popular entrées include meat items such as beef, pork, and poultry. Seafood alternatives are also popular, as are entrée salads and vegetarian items that do not contain animal products.

  Many entrées are items served hot, but others, such as a chef's salad, are served cold. Some entrées, such as a salad served with grilled chicken, may be hot and cold. Casseroles are entrées that can include meat or seafood with pasta, cheese, and other ingredients, usually baked in a deep dish.
- **House specialties:** Some operations specialize in items such as barbecued ribs or steaks and feature them in a separate menu classification.
- **Appetizers (starters):** Appetizers are served as the first course in a meal. They may be hot such as stuffed mushrooms, or cold such as shrimp cocktail.
- **Soups:** Some menus present soups separately, while others include them with appetizers or salads. Most soups are served hot and can be clear (broth) or thick (chowder) and served with meats, seafood, or poultry. Some soups may be served cold. For example, vichyssoise is a cold soup made from potatos, onions, and leeks.
- **Salads:** Entrée salads and accompaniment (side) salads may be offered. Many are made from lettuce, leafy greens, and other vegetables. Salads can also be made from fruits.

- **Sandwiches and wraps:** Common sandwiches include hamburgers, BLTs (bacon, lettuce, and tomato sandwiches), and deli favorites such as ham and turkey. Sandwich wraps include tortillas and pita bread made with numerous types of fillings.
- **Vegetables and accompaniments (side dishes):** Potatoes and other vegetables are available in many preparation styles. Fresh vegetables are an important part of many menus. Other dishes include rice and pasta. Accompaniments are often suggested by the operation's concept, such as pasta items in a property with an Italian theme.
- **Desserts:** After-dinner items are often included on menus. A separate dessert menu is sometimes used, or a dessert cart or tray can be brought tableside.
- **Beverages:** Traditional items such as coffees, teas, and soft drinks may be listed on the menu. Sometimes separate coffee menus are used. Some menus suggest traditional cocktails, and many operations feature alcoholic beverages. These may be offered on separate menus. Many properties have beer and wine lists.

How many menu classifications are needed? Planners must examine these issues:

- How many categories are needed for the variety of menu items to be offered?
- Is a category needed to list unique items?
- Does the market want the same basic types of menu items? If so, fewer classifications are needed.
- What are the most frequent reasons for customer visits? Operations with a high-volume lunch business serve customers more quickly when fewer classifications are available. A meal with several courses may be preferred for dinner customers.
- Will the menu be easy to read? Too few classifications create long lists in each category. On the other hand, a menu classification with just one or two items may not be appropriate unless it is a specialty category.

Step 4: Consider Potential Menu Items

One way to select items for each menu classification is to consider many items and then decide which to eliminate. When this strategy is used in an operation that already has a menu, the following procedure can be followed:

- Items within a classification are eliminated if they are unpopular, unprofitable, contain ingredients that are difficult to purchase, or are causing production problems.

- A list of possible items to add is suggested by menu planners. Trends may be considered, and the dining-room manager can remind the team what customers are requesting.
- The list of possible menu items is reduced to a manageable size by considering factors discussed later in this section.

When a menu is being planned for a new establishment, the same basic elimination process can be used. A relatively long list of possible menu items for each classification is developed, and items not meeting the property's selection concerns are eliminated.

Some special menu planning tools can be helpful:

- Copies of menus can help planners think about items that have, and have not, been popular, profitable, or easy to produce and serve.
- Copies of competitors' menus may suggest items that should be added.
- Menu evaluation information about item profitability and popularity is useful. This topic is discussed later in this chapter.
- Standardized recipes for possible menu additions are needed to ensure that items suggested can be produced to required standards. A **standardized recipe** is the set of instructions to produce and serve a food or beverage item that will help ensure that quality and quantity standards will be consistently met. If standards can be met, planners can consider adding the new item to the menu.
- Product inventory and ingredient availability reports are important. These reports show availability of products and ingredients of the right quality and acceptable cost during the time for which the menu will be available. This information might be provided by suppliers.
- Input from managers, employees, and customers. Each can provide input about current items and make suggestions about proposed items.

Step 5: Select Specific Menu Items

Menu items needed for each classification are often planned in the following order:

1. Entrées
2. Appetizers
3. Soups and salads
4. Sandwiches and wraps
5. Vegetables and accompaniments
6. Desserts
7. Beverages

Entrées are selected before items in other classifications because customer decisions about where to eat are often based on entrée selections. Also, operations with dining themes need specific types of entrées to help "deliver" their cuisines to customers.

Several factors should be considered when selecting menu items:

- **Variety:** A range of serving temperatures, preparation methods, textures, shapes, and colors should typically be available within each menu classification. Too little variety is generally not good because even persons in a closely defined market like different items. Too many items may cause customers to take longer to order, and more time and effort will be needed to purchase, store, and produce the items.
- **Temperature:** Customers typically expect some items to be served hot like mashed potatoes, and other items to be served chilled such as tossed salad. Temperature is also a very important food safety concern. Can required temperatures be consistently maintained? If there are any doubts about food safety issues, affected items should not be placed on the menu.
- **Nutrition:** Many customers are concerned about nutrition. Some operations may offer a specific menu section devoted to healthy dining options. Other establishments may not break out these entrées, but instead simply elect to make some items more appealing to health-conscious consumers with minor changes. Deep-fried chicken can be offered baked without the skin, and entrées with a rich sauce can be broiled and offered without a sauce or with sauce on the side. Smaller servings of some menu items can be offered. When these strategies are used, existing menu items will appeal to all customers with few production challenges.
- **Texture:** Texture relates to how a food feels. Menu items can be soft or hard, firm or crunchy, liquid or solid, wet or dry. Many customers like cooked vegetables such as carrots to be "crisp-tender" rather than soft.
- **Shape and size:** Menu items can be round, square, or long, and they can be served in a flat or tall portion. Sometimes a garnish, such as slivers of tall, thin vegetable strips, can impact perception of an item's height. A **garnish** is an edible decoration used to make a menu item attractive.
- **Flavor:** Flavors relate to taste. Common tastes are sweet, sour, salty, and bitter. Customers also use terms such as *hot* (buffalo wings), *spicy* (Thai dishes), and *smoked* (barbecued meats) to describe taste. Professional chefs know that flavor really refers to the total experience one has with food as it is consumed. All senses including sight, taste, touch, smell, and even the sound of a sizzling steak are involved in a food's flavor.
- **Color:** Color is part of a menu item's appeal. Multiple colors are preferred; it is best not to serve fried chicken, a baked potato, and baked beans on the same plate because all of these items are

brown. Preparation methods may affect color. For example, the coating of fried chicken is typically darker than the surface of oven-baked chicken. Color also relates to the specific food item. For example, customers expect broccoli to be a bright, medium-dark green.

- **Composition and balance:** Sometimes managers have no control over how food items relate to each other. For example, buffet customers select the food items they want. However, when items are plated for service, the way food items look in relation to each other should be considered (*Exhibit 1.6*).
- **Possible selling price:** The cost of ingredients is an important factor in the price charged for the item. Planners must consider how much customers will pay and still receive a perceived value for their purchase.
- **Test results:** Proposed menu items should be prepared according to the standardized recipe and then taste-tested. Items can be sampled by employees and customers. Feedback can be provided orally or by use of a comment card. Positive feedback is a very important factor in the decision to offer an item.

Exhibit 1.6

The last two steps in menu planning relate to developing menu drafts and revising the menu as necessary.

MENU DESIGN

Menu planners must consider basic information before designing the menus for their properties. In particular, they must consider what type of menu they plan to offer, and the length of the menu.

Background Information

There are several common types of menus:

- **À la carte menu:** These menus have different prices for each menu item. The customer's charge is based on the prices of the items ordered.
- **Table d'hôte menu:** These menus offer an entire meal at a set price. Examples include banquet and buffet menus. Some establishments also set a price for an entire meal with a specified number of courses.
- **Cyclical (cycle) menu:** These menus are planned for a specified time period and then repeated. For example, a cafeteria may offer a 28-day menu that is repeated every four weeks.
- **Du jour menu:** A du jour menu changes daily. Many establishments offer daily specials in addition to their regular menu items.

THINK ABOUT IT . . .

French culinary terms are widely used in modern operations. Examples include *toque* (chef's hat), *sous chef* (chef's assistant), *bouillon* (soup from beef stock), and *haute cuisine* (high-quality food). What other French terms are in common use?

Menu planners consider length in early drafts of menu design. Length is partially determined when menu items are selected because the menu must inform customers about all available items. The focus on length reemphasizes the importance of carefully considering the target market's preferences as menus are planned.

Longer menus require more time for customer review and raise the possibility of more questions. More review and more questions slow the ordering process. While customers should be allowed the time they wish to dine, menus designed to sell recognize that table turns allow more customers to be served and additional revenues to be generated. **Table turn** refers to the number of times a dining-room table is occupied during a meal period.

Menu Design Procedures

Exhibit 1.4 indicated that developing a menu draft (step 6) and revising the menu as needed (step 7) are two final menu planning steps. Numerous menu design strategies are important:

- The menu must be attractive and present a good first impression. Managers should ask themselves if they are proud of their menu. Does it represent the operation to customers in the most favorable way?
- The menu should not be cluttered; generally, approximately one-half of the space on a menu page should be blank.
- All words must be spelled correctly, and avoid using any hard-to-understand foreign terms.
- Menu descriptions, not selling prices, should be the focus of attention.
- The menu's size should allow it to be comfortably read and handled at the table.
- Menus should be durable and easy to clean unless they are single-use.

Modern menus are important merchandising tools that emphasize specific menu items to encourage customers to select them. Menus must be thoughtfully developed to effectively communicate with customers and influence sales.

Menus that promote specific items use space wisely. Unique items need more space for description. An item with a name like "Chef Craig's Garden Soup" must be described in detail. If it is not, servers will require more time to explain the soup to customers. Simple ways to promote desired menu items include putting them in a box or using shading or a different type style. Use of colored print and color photos of selected menu items are other possibilities.

Manager's Memo

Good menu planners avoid common design problems. They know that a small type size is hard to read, especially in dimly lighted areas. They do not use decorative typefaces that are difficult to read.

They design menus to sell items that are popular and profitable, and they ensure that menu descriptions accurately represent the items that will be served. They know that a good design allows customers' eyes to move easily from category to category and from item to item.

LOCATION OF MENU ITEMS

Menu layout relates to where items are placed. Normally, menu item classifications influence the layout. If appetizer, soup and salad, and entrée classifications have been identified, the layout must include space for these classifications.

Some menu designers place categories on the menu in the sequence of service. For example, appetizers are listed before soups and salads, which are listed before entrées. However, menus have "prime real estate" areas most frequently viewed by customers. These areas should contain the items planners most want to sell. *Exhibit 1.7* shows prime real estate areas for some common menus.

Exhibit 1.7

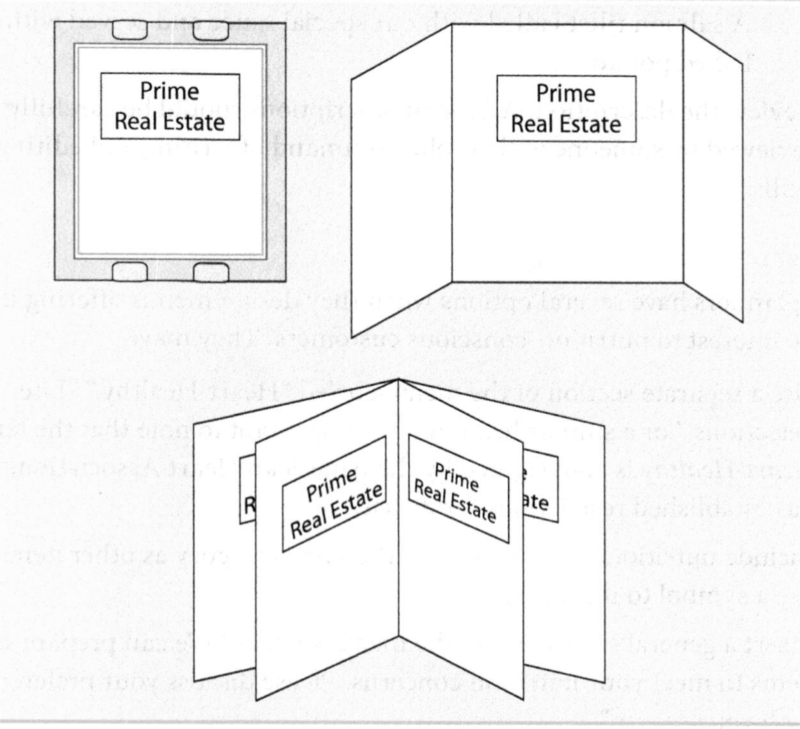

MENU ITEM DESCRIPTIONS

Menu item descriptions answer potential questions of menu readers. For example, what should customers know about Seafood Fettuccine? Many customers may know that fettuccine is a pasta. However, what type of seafood and sauce is used in preparation? Customers may become annoyed if they must ask their server, and servers will have less time to provide the quality of service that other customers deserve.

Menu planners must consider these issues when writing menu descriptions:

- Write plainly. Avoid technical culinary terms and foreign words unless they are menu names on menus featuring food from different countries or cuisines. For example, consider writing *dumplings* instead of *quenelles* and *finely shredded vegetables and herbs* instead of *chiffonade.*

- Define menu items carefully and correctly. For example, white New England clam chowder is not the same as red Manhattan clam chowder.
- Spell all words correctly, such as *Portobello* mushrooms, *vinaigrette* dressing, and *au jus*.
- Write carefully. For example, only the last description is correct in the following list:
  - A salmon fillet served with a baked potato ladled with our special sauce.
  - A salmon fillet ladled with our special sauce and baked potato.
  - A salmon fillet ladled with our special sauce and served with a baked potato.
- Review the description. All menu descriptions should be carefully reviewed by someone with a solid command of writing and editing skills.

NUTRITION AND MENU DESIGN

Menu planners have several options when they design menus offering items of specific interest to nutrition-conscious customers. They may:

- Use a separate section of the menu labeled "Heart-Healthy," "Lite Selections," or a similar heading. It is important to note that the term *Heart-Healthy* is copyrighted by the American Heart Association, which has established restrictions for its use.
- Include nutritional alternatives in the same category as other items, and use a symbol to indicate them.
- Insert a general statement on the menu, such as "We can prepare some items to meet your nutrition concerns. Please discuss your preferences with your server."

Menu nutrition content and health claims must be truthful. It is not appropriate and may be unlawful to misstate dietary claims in menu descriptions.

Terms such as *fat-free, low-fat,* and *lean beef* have specific technical meanings. It is typically better to state the ingredients an item is made with, such as margarine or low-fat cottage cheese. Nutrition labeling requirements are affected by federal, state, and local requirements, and they may vary depending on a number of factors such as the number of units in a chain.

MENU ACCURACY

Those writing menu descriptions must accurately describe the items being served, and they must know about truth-in-menu issues. Truth-in-menu requirements, and laws in some locations, dictate that menu descriptions honestly tell the quantity, quality, point of origin, and other information so menu readers will be fully informed about the items they order.

Here are some examples of menu accuracy:

- **Quantity:** A two-egg omelet should contain two eggs.
- **Quality:** The term *prime* used to describe a steak refers to a specific U.S. Department of Agriculture (USDA) grading standard. Only the quality of product actually used should be indicated on the menu.
- **Price:** If there are extra charges (for example, for a call brand liquor), the price should be identified. A **call brand** is a specific brand of liquor requested by a customer.
- **Brand name:** If a specific product brand such as Coca-Cola® or Pepsi® is noted on the menu, this brand should be served.
- **Product identification:** Maple syrup and maple-flavored syrup, for example, are not the same.
- **Preservation:** Frozen green beans should not be called "fresh."
- **Food preparation:** "Made on site" does not apply to a convenience food product produced elsewhere.
- **Verbal and visual presentation:** A menu photograph depicting eight shrimp on a shrimp platter means that eight shrimp should be served.
- **Dietary and nutritional claims:** If the menu indicates "egg substitutes are available," they must be used when ordered.
- **Preparation style:** The preparation method listed on the menu should be used. If a grilled item is listed, the item cannot just have mechanically produced grill marks and then be steamed before service.

Caution statements should also be added to menus to alert customers about potential problems such as small bones in fish and eating undercooked meats as shown in *Exhibit 1.8*.

Exhibit 1.8

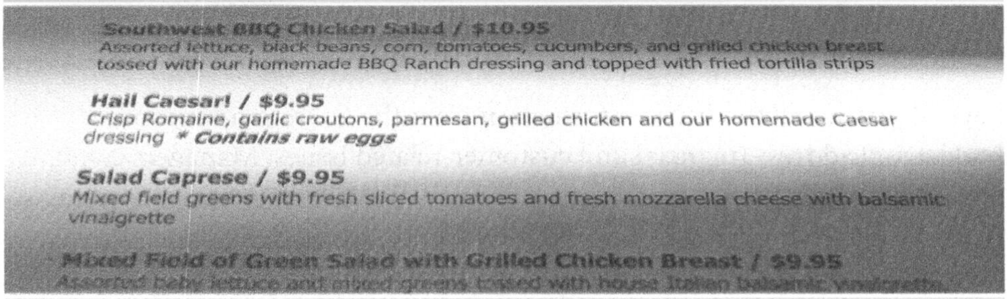

MENU APPEARANCE AND CONSTRUCTION

The menu cover provides an early visual impression of the establishment and the dining experience. The front cover design should fit with the establishment's décor and theme and should include its name. Padded menu covers made from imitation or genuine leather, velvet, or other materials are available. Some operations use desktop publishing systems to print and change menus frequently. A desktop publishing system uses a personal computer and specific software to create high-quality page layouts that resemble those developed by a professional printer.

A menu can be small (8.5 by 11 inches or smaller) or large (15 by 20 inches or larger). If a menu panel is too small, however, the print is typically small and can be hard to read. A menu that is too large may be inconvenient to handle.

A menu's construction material affects its durability and ability to be cleaned. If a dirty menu cannot be cleaned, it must be discarded regardless of cost.

An otherwise attractive menu is affected by the menu's type style. The primary concern in choosing a menu typeface is its readability. Type that is too small or too closely spaced can create confusion. The lighting level in the dining area affects the menu's readability. Wise menu designers look at their menus in the dining room rather than just in a well-lighted office.

Subheadings are sometimes used within menu item classifications to separate, for example, seafood, beef, chicken, and pork entrées or other items that can be broken into groups.

Darker colors on a light background, or vice versa, can help emphasize selected items. Decorative details such as pictures, drawings, and designs can add appeal. A menu's graphic design can "bring it all together" and help ensure that a well-designed menu is also attractive and readable.

> **Manager's Memo**
>
> Space is needed for menu item classifications, but it is also required for other purposes:
>
> - **General descriptions:** What accompaniments go with the entrées? What makes the salad bar especially inviting to customers?
> - **Cross-selling opportunities:** **Cross-selling** relates to strategies used to advertise products and services offered in addition to those noted on a specific menu. For example, a dinner menu can promote a weekend brunch, describe banquet capabilities, or advertise lunch hours.
> - **Property information:** Name, address, telephone number, and hours of operation are among the information that should be included on take-home menus.
> - **Special information:** Examples may include a note about service charges for groups of a specified size or the establishment's history and mission statement.

MENU ITEM SELLING PRICES

Menu planners must consider the selling price of items when they plan and design their menus. The selling price should help the operation meet its financial goals while providing value to customers.

Objective Menu Pricing Methods

Several pricing methods use information from the operation's approved budget and address financial and customer-related issues. Managers can choose from the food cost percentage method, contribution margin pricing method, ratio pricing method, or prime cost pricing method, depending on the financial and accounting needs of their organization.

FOOD COST PERCENTAGE METHOD

The **food cost percentage** is the percentage of the revenue that goes to the cost of purchasing food.

The formula to calculate a food cost percentage is

Food cost ÷ Selling price = Food cost percentage

If a menu item had a food cost of $5.85 and a selling price of $18.95, then a manager can use the formula to calculate the item's food cost percentage:

$5.85	÷	**$18.95**	=	**31%**
Food cost		**Selling price**		**Food cost (rounded)**

Managers can use the food cost percentage to arrive at menu selling prices. The food cost percentage pricing method goes by several names including the *simple markup method* and the *factor method*. While the different names may sound confusing, this method is actually the simplest of all menu pricing methods. The reason is because it is based only on the cost of food required to make a menu item.

If a manager is pricing an item with the food cost percentage pricing method, he or she can use the cost of the food and the food cost percentage to arrive at a suggested menu price. Assume that a manager needs to achieve a 31 percent food cost percentage and that a particular item has a food cost of $5.85. The manager would calculate the suggested menu price using the following formula:

Food cost ÷ Food cost percentage = Selling price

This is calculated as follows:

$5.85	÷	**0.31**	=	**$18.90**
Food cost		**Food cost percentage**		**Selling price**

By applying basic algebra, the use of the formula means that if any two of the numbers in the formula are known by a manager, the third number can easily be calculated. Thus:

If food cost and selling price are known, the formula to compute *food cost percentage* is as follows:

Food cost ÷ Selling price = Food cost percentage

If food cost and targeted food cost percentage are known, the formula to compute *selling price* is as follows:

Food cost ÷ Food cost percentage = Suggested menu price

If targeted food cost percentage and desired selling price are known, this is the formula to compute *food cost*:

Selling price × Food cost percentage = Food cost

OPEN FOR BUSINESS

RESTAURANT TECHNOLOGY

Technology provides new options for menu design in quick-service restaurants (QSRs). Everyone is familiar with the menu boards that are placed at order counters and the colorful and sometimes overcrowded signs in drive-through lanes.

Consider the possibility of electronic menu boards and display units that can be programmed to change menu items, photos, selling prices, and related information at certain times of day. What about streaming video with suggestive selling advertisements or flashing displays that draw attention to certain items? These technologies are currently available and may be the future of QSR operations, especially in large chain organizations.

Self-order kiosks are likely to be increasingly used in QSR properties. Customers will go to a touch-screen unit to place their order. They can respond yes or no to questions like "Want some fries with your order?" These kiosks will also calculate and accept payments.

OPEN FOR BUSINESS

MANAGER'S MATH

The manager is calculating proposed menu prices for three new menu items. Using the information, calculate the selling price he or she should charge for each new menu item and then use the information to answer the questions that follow.

Item	Food Cost	Target Food Cost Percentage	Selling Price
Sirloin Steak	$ 6.50	38%	
Baked Chicken	$ 3.27	23	
Lake Trout	$ 4.71	27	

1. What is the selling price for Sirloin Steak resulting from the formula's calculation?
2. What is the selling price for Baked Chicken resulting from the formula's calculation?
3. What is the selling price for Lake Trout resulting from the formula's calculation?
4. If you were the manager, would you adjust any of these prices up or down prior to printing them on the menu? Why?

(Answers: 1. $17.11; 2. $14.22; 3. $17.44)

In the previous example, if the manager wanted to sell an item for $18.95 and knew that the targeted food cost was 31 percent, then the allowable food cost would be calculated this way:

$18.95	×	**0.31**	=	**$5.85**
Selling price		**Food cost percentage**		**Food cost (rounded)**

After mastering the basic formulas, the food cost percentage pricing method is simple to use. However, the selling price that results from using this method is often simply a starting point. To illustrate, assume an item has a food cost of $5.11 to make. The manager has a 32 percent target food cost. Using the food cost percentage formula yields this selling price:

$5.11	÷	**0.32**	=	**$15.97**
Food cost		**Food cost percentage**		**Selling price**

In this case, the managers would likely round the selling price to $15.99 or another less awkward selling price.

Be aware that applying the same division or multiplication factors to all menu items may not be in the operation's best interest. Some managers feel the food cost percentage pricing method tends to overprice high food cost items and underprice low food cost items such as soups, pasta, chicken, beverages, and desserts. An additional weakness is that the food cost percentage pricing approach ignores the cost of labor. For example, a menu item that takes a long time to prepare will have a higher labor cost than an item that is prepared quickly. If the food costs of the two items are identical they would both have the same selling price if the food cost percentage pricing method is used.

Market factors, what potential customers are willing to pay for specific menu items, and what the competition is charging should all influence pricing decisions. Prices should be adjusted up or down based on psychological pricing considerations and other subjective factors. These adjustments may result in the selling prices for many menu items being set at a food cost percentage higher than management might desire, but there will be others with a lower than average food cost percentage. However, the goal is for the total menu to average out to a targeted food cost percentage.

CONTRIBUTION MARGIN PRICING METHOD

A menu item's contribution margin (CM) is found by subtracting its food cost from its selling price. The CM indicates how much is left to "contribute" to nonfood costs (every cost except food) and profit.

CM pricing involves working this process in reverse. Planners must determine the total of nonfood costs and profit and the portion to be generated by a single sale. Assume that the approved operating budget indicates that all nonfood costs will be $315,000, the profit goal is $31,000, and 70,000 customers are expected to be served.

Step 1. Calculate the average CM per customer:

(Nonfood costs + Profit) ÷ No. of customers = Average CM per customer

In this example:

($315,000	+	$31,000)	÷	70,000	=	$4.94
Nonfood costs		Profit		No. of customers		Average CM per customer

Step 2. Determine the selling price for the menu item by adding the food cost to the CM.

For example, the base selling price for a menu item with a $5.20 food cost would be:

$5.20	+	$4.94	=	$10.14
Food cost		CM		Base selling price

CM menu pricing is easy because the necessary information is in the operating budget. It is practical when the nonfood costs required to serve each customer are basically the same. This method reduces the range of selling prices because the only difference is the cost of the food item selected.

RATIO PRICING METHOD

The ratio pricing method considers the relationship or ratio between food costs and CM (nonfood costs and profit) and then uses the ratio to develop base selling prices.

Assume that the operating budget indicates food costs of $310,000, nonfood costs of $620,000, and a profit goal of $48,000.

Step 1. Calculate the ratio of food cost to nonfood costs and profit (the CM):

CM ÷ Food cost = Ratio

In this example:

($620,000	+	$48,000)	÷	$310,000	=	2.15
Nonfood costs		Profit		Food cost		Ratio

Now the manager knows that for each dollar of revenue to pay for food costs, an extra $2.15 in revenue must be generated for CM requirements.

Manager's Memo

The food cost percentage method is often referred to as the *markup method* or *factor method* because managers can use a factor as a shortcut to calculate selling prices. Dividing any targeted food cost percentage into 1.00 yields a factor that can be multiplied by an item's food cost to yield its selling price. The following factor table shows the factors that result for several popular food cost percentage targets.

Target Food Cost %	Factor
20	5.00
25	4.00
28	3.57
30	3.33
32	3.12
34	2.94
36	2.77
40	2.50
42	2.38

Assume that the food cost of the item is $1.74 and the targeted food cost percentage is 32 percent. Using the factor from the table, the selling price would be calculated as follows:

Cost of food × Factor = Selling price

Or,

$1.74	×	3.12	=	$ 5.43
Cost of food		Factor		Selling price

Step 2. Calculate the CM amount for the menu item.

Multiply the item's food cost by the ratio:

$$\textbf{Food cost} \times \textbf{Ratio} = \textbf{CM}$$

In this example, assume that the food cost is $3.78. The formula applied would look like this:

$$\underset{\text{Food cost}}{\$3.78} \times \underset{\text{Ratio}}{2.15} = \underset{\text{CM}}{\$8.13}$$

Step 3. Determine the menu item's selling price by adding the CM to the menu item's food cost.

$$\textbf{Food cost} + \textbf{CM} = \textbf{Menu price}$$

In this example:

$$\underset{\text{Food cost}}{\$3.78} + \underset{\text{Required CM}}{\$8.13} = \underset{\text{Menu price}}{\$11.91}$$

The ratio pricing method is simple because it is based on budget information. It has one challenge: An operation selling food and alcoholic beverages must separate nonfood costs and profit requirements between the two revenue centers to determine the required food-related financial information.

PRIME COST PRICING METHOD

In the restaurant and foodservice industry, food costs and labor costs together are considered **prime costs**. The prime cost pricing method requires managers to consider the labor cost required to make a menu item as well as the item's food cost. The formula used in prime cost pricing is as follows:

$$\left(\begin{matrix}\text{Labor}\\\text{cost}\end{matrix} + \begin{matrix}\text{Food}\\\text{cost}\end{matrix}\right) \div \left(\begin{matrix}\text{Target labor}\\\text{cost percentage}\end{matrix} + \begin{matrix}\text{Target food}\\\text{cost percentage}\end{matrix}\right) = \begin{matrix}\text{Selling}\\\text{price}\end{matrix}$$

Use of the prime cost pricing method requires management to determine the cost of direct labor spent on preparing an item. This amount is added to food cost to determine the prime cost. For example, assume that the food cost for an item is $4.15 and the labor cost is $1.75. The item's prime cost would be $5.90.

A manager derives the selling price based on prime cost by first establishing a combined labor and food cost percentage. This figure is called a prime cost percentage. If the manager desires a 20 percent labor cost and a 33 percent food cost, the prime cost percentage would be 53 percent:

$$\underset{\text{Labor cost}}{20\%} + \underset{\text{Food cost}}{33\%} = \underset{\text{Prime cost}}{53\%}$$

In this example, the selling price formula would be as follows:

($1.75	**+**	**$4.15)**	**÷**	**(20%**	**+**	**33%)**	**=**	**$11.15**
Labor cost		**Food cost**		**Target labor cost**		**Target food cost**		**Selling price**

Or,

$5.90	**÷**	**53%**	**=**	**$11.15**
Prime cost		**Prime cost percentage**		**Selling price (rounded)**

The item with a $5.90 prime cost would be divided by 53 prime cost percent to arrive at an $11.15 menu selling price.

Managers tend to use the prime cost pricing method if menu items vary widely in their preparation time and costs. This pricing method allows for the true cost of the labor to be associated with each individual menu item.

MORE ABOUT OBJECTIVE MENU PRICING

Objective pricing methods are easy to use because the numbers are in the operating budget. The most challenging calculation involves determining the food cost for each menu item. The manager must have standardized recipes costed with current financial information to use any pricing method based on food costs. Costing considers the cost to produce all or a single serving of a recipe by calculating ingredient costs and the number of servings the recipe yields.

Fortunately, standardized recipes should be available because they are needed for other aspects of operating control, and recipe costing has become much easier and faster with the use of applicable software.

ANALYZING SALES MIX: MENU ENGINEERING

A **sales mix analysis**, also called menu engineering, is a study designed to determine the popularity and profitability of competing items on a menu. A major purpose of the analysis is to monitor the effectiveness of menu items to best meet the needs of customers and maximize the profits for the operation. The results of the analysis determine whether changes in menu pricing and menu design are needed.

Sales mix analysis involves determining which menu items are most popular and which contribute the most money to expenses and profit. Popularity relates to number of sales of each menu item, and profitability is based on CM (Revenue – Food cost).

Manager's Memo

Sales mix analysis focuses a manager's attention on a menu item's food cost, since CM is what remains after food costs are deducted.

Labor cost is another very important concern for managers. As they make decisions based on sales mix analyses, they should also think about the labor costs required to produce the items. Some menu items are very labor-intensive, requiring extra work to produce them. This will result in a higher-than-average labor cost for those items.

Labor costs should be considered when item profitability is assessed because high labor costs, not directly considered in the sales mix analysis, reduce the actual profitability of a menu item.

Performing a Sales Mix Analysis

Several steps are involved in performing a sales mix analysis:

1. Select items to compare.
2. Determine menu mix percentage (popularity rate).
3. Compare menu items and menu mix popularity percentage.
4. Determine menu item CMs.
5. Determine average CM (profitability).
6. Compare menu item and average CMs.
7. Classify menu items.

The sales mix analysis is done using a format similar to *Exhibit 1.9*. While it suggests a manual (hand) tally of the information, software is also available.

Step 1: Select Items to Compare

A sales mix analysis is done on competing items in a single menu classification, such as entrées.

The first step is to list the menu items to be analyzed in column A of the sales mix analysis worksheet. *Exhibit 1.9* lists the entrées at Union Street Cafe, taken directly from its menu.

Exhibit 1.9

SALES MIX ANALYSIS FOR UNION STREET CAFE

A Menu Item	B Number Sold	C Menu Mix % (B ÷ ΣB)	D Selling Price	E Item Food Cost	F Item Contribution Margin (D − E)	G Total Item CM (B × F)	H CM Category	I MM % Category	J Menu Item Classification
Skewered Shrimp	370	19.6%	$12.95	$4.51	$8.44	$3,122.80	Low	High	Plow horse
Pork Medallions	250	13.3	15.50	5.71	9.79	2,447.50	High	High	Star
Rib-Eye Steak	420	22.2	14.95	5.38	9.57	4,019.40	High	High	Star
Herbed Chicken	290	15.3	12.95	3.37	9.58	2,778.20	High	High	Star
Planked Salmon	140	7.4	15.95	5.90	10.05	1,407.00	High	Low	Puzzle
Beef Stroganoff	210	11.1	11.50	4.18	7.32	1,537.20	Low	High	Plow horse
Veal Piccata	100	5.3	13.50	7.15	6.35	635.00	Low	Low	Dog
Filet Oscar	110	5.8	15.95	7.82	8.13	894.30	Low	Low	Dog
Total Σ	**1,890**					**$16,841.40**			

Step 2: Determine Menu Mix Percentage (Popularity Rate)

In column B, list the number of each entrée sold during the time period covered by the analysis. Then calculate the total by adding the individual sales. Note that 1890 entrées were sold, as shown in column B.

Now the **menu mix percentage** for each entrée can be calculated by dividing the number of each specific item sold by the total number of items sold. For example, Skewered Shrimp (line 1) represents 19.6 percent of all entrées sold:

370	÷	**1,890**	=	**19.6%**
No. of skewered shrimp entrées sold		**Total entrées sold**		**Menu mix %**

Menu mix percentages (abbreviated MM %) for each item are recorded in column C.

The sales mix analysis model assumes that a menu item is popular if it sells at least 70 percent of the expected sales for a menu item. This is calculated by dividing 100 percent by the number of menu items being analyzed and then multiplying that number by 70 percent. The result is the **menu mix popularity percentage**: the percentage of total menu items that must be sold for a menu item to be considered popular when sales mix analysis is performed.

For example, Union Street Cafe features eight entrées (column A). Therefore, the menu mix popularity percentage (MM %) is 8.75 percent:

100%	÷	**8**	=	**12.5%**
Total sales		**No. of different entrées**		**Expected sales mix %**

12.5%	×	**70%**	=	**8.75%**
Expected sales mix %		**Allowable sales %**		**Menu mix popularity %**

Step 3: Compare Menu Items and Menu Mix Popularity Percentage

After the menu mix popularity percentage is calculated, the MM percentage of each menu item is compared to this benchmark to determine whether the item is popular or unpopular.

- Items with an MM percentage at or above the menu mix popularity percentage are considered popular sellers.
- Items with an MM percentage below the menu mix popularity percentage are considered unpopular.

OPEN FOR BUSINESS

MANAGER'S MATH

The menu mix popularity percentage is the baseline against which all menu items are compared during sales mix analysis to determine popularity.

What is the menu mix popularity percentage in the following cases?

- The menu has 25 competing items.
- The menu has 12 competing items.
- The menu has 6 competing items.

(Answers: 2.80%; 5.83%; 11.67%)

Manager's Memo

The food cost for a menu item is calculated by costing the standardized recipe for each item included with the meal and calculating its cost per serving. Column E in *Exhibit 1.9* indicates that the food cost for Skewered Shrimp is $4.51. Assume that a customer ordering this item will receive the entrée, a salad with dressing, a potato choice, and a bread roll with butter. The food cost for Skewered Shrimp is the total per-portion cost for all items the customer receives when he or she orders that entrée.

If an item is considered popular, *High* is recorded in the MM % Category (column I). If an item is considered unpopular, *Low* is recorded.

In this example, the MM percentage of Skewered Shrimp is 19.6 percent (column C). This is significantly higher than the MM popularity percentage of 8.75 percent, so the popularity of Skewered Shrimp is high (column I).

The menu mix percentage category (abbreviated MM % Category) of each item is shown in column I. The comparison process used to reach these results is shown in *Exhibit 1.10*.

Exhibit 1.10

MENU MIX PERCENTAGE CATEGORY RESULTS

A Menu Item	B Number Sold	Menu Mix Popularity %	C Menu Mix %	I MM % Category
Skewered Shrimp	370	8.75%	19.6%	High
Pork Medallions	250	8.75	13.3	High
Rib-Eye Steak	420	8.75	22.2	High
Herbed Chicken	290	8.75	15.3	High
Planked Salmon	140	8.75	7.4	Low
Beef Stroganoff	210	8.75	11.1	High
Veal Piccata	100	8.75	5.3	Low
Filet Oscar	110	8.75	5.8	Low
Total	1,890			

Step 4: Determine Menu Item Contribution Margins

The fourth step in sales mix analysis is to determine each menu item's CM. Start by listing the selling price of each entrée in column D (see *Exhibit 1.9*). For example, the menu price for Skewered Shrimp is $12.95.

Next, list the food cost of each item in column E. Note that the food cost for Skewered Shrimp is $4.51.

Recall that a menu item's CM is the difference between the item's revenue (selling price) and its food cost.

The CM for each entrée at Union Street Cafe is shown in column F of *Exhibit 1.9*. Note that the CM from the sale of one serving of Skewered Shrimp is $8.44. It is calculated as follows:

$12.95	–	$4.51	=	$8.44
Item selling price (col. D)		Item food cost (col. E)		Item contribution margin (col. F)

The CM from total sales of each menu item is determined by multiplying the number of items sold (column B) by the item's CM (column F). The total CM from the sale of all 370 servings of Skewered Shrimp is shown in column G:

$8.44	×	**370**	=	**$3,122.80**
Item contribution margin (col. F)		**No. of items sold (col. B)**		**Total item contribution margin (col. G)**

The total CM for all entrées sold at Union Street Cafe was $16,841.40, shown at the bottom of column G.

Step 5: Determine Average Contribution Margin (Profitability)

Each entrées' average CM is determined by dividing the total CM (bottom of column G) by the total number of entrées sold (bottom of column B):

$16,841.40	÷	**1,890**	=	**$8.91**
Total contribution margin (col. G)		**Total no. of items sold (col. B)**		**Average contribution margin**

Step 6: Compare Menu Item and Average Contribution Margins

The CM of each menu item should be compared to the average CM to determine the item's profitability. If the item's CM is higher than the average, then *High* is recorded in column H. If the item's CM is lower than the average, *Low* is recorded.

In *Exhibit 1.9*, notice that the CM of $8.44 for Skewered Shrimp (column F) is lower than the average CM of $8.91. Therefore, it has a low CM.

The contribution margin for each item in the example is shown in column F of *Exhibit 1.9*. The comparison process is shown in *Exhibit 1.11*.

OPEN FOR BUSINESS

MANAGER'S MATH

The average CM is the benchmark against which all menu items are compared in sales mix analysis to determine whether each item is profitable.

- What is the average CM when the sale of 2,410 menu items yields a total CM of $21,931?
- What is the average CM when the sale of 3,250 menu items yields a total CM of $52,975?
- What is the average CM when the sale of 2,110 menu items yields a total CM of $15,980?

(Answers: $9.10; $16.30; $7.57)

Exhibit 1.11

CONTRIBUTION MARGIN CATEGORY RESULTS

A Menu Item	B Number Sold	F Item CM	Average CM	H CM Category
Skewered Shrimp	370	$ 8.44	$8.91	Low
Pork Medallions	250	9.79	8.91	High
Rib-Eye Steak	420	9.57	8.91	High
Herbed Chicken	290	9.58	8.91	High
Planked Salmon	140	10.05	8.91	High
Beef Stroganoff	210	7.32	8.91	Low
Veal Piccata	100	6.35	8.91	Low
Filet Oscar	110	8.13	8.91	Low
Total	**1,890**			

Step 7: Classify Menu Items

The last step in sales mix analysis is to classify the menu items based on a combination of their MM percentage category (column I of *Exhibit 1.9*) and their CM category (column H). The following names are used in the classification system:

- **Star:** This is an item with a high MM percentage and a high CM.
- **Plow horse:** A plow horse is an item with a high MM percentage and a low CM.
- **Puzzle:** A puzzle is an item with a low MM percentage and a high CM.
- **Dog:** This is an item with a low MM percentage and a low CM.

Menu improvements are based on these classifications. The results of assigning menu item classifications to the entrées at Union Street Cafe are shown in column J of *Exhibit 1.12*.

Exhibit 1.12

MENU ITEM CLASSIFICATION RESULTS

A Menu Item	H CM Category	I MM % Category	J Menu Item Classification
Skewered Shrimp	Low	High	Plow horse
Pork Medallions	High	High	Star
Rib-Eye Steak	High	High	Star
Herbed Chicken	High	High	Star
Planked Salmon	High	Low	Puzzle
Beef Stroganoff	Low	High	Plow horse
Veal Piccata	Low	Low	Dog
Filet Oscar	Low	Low	Dog

Sales Mix Analysis and Menu Change

The purpose of learning each menu item's popularity and profitability is to determine how to merchandise the items on the menu. Managers can use many strategies to improve menu items. However, it is important to make changes in line with the brand image, competitors, production and service considerations, and customer expectations.

STARS

Stars are menu items that are profitable and popular. The following strategies can be used to manage stars:

- Do nothing, as these items already sell well and make money for the operation.
- Maintain quality standards.
- Ensure that these items are in highly visible menu locations.
- Promote these items aggressively with table tents and **suggestive selling**: the tactic of using recommendations to ensure that customers know about the products and services offered by the restaurant or foodservice operation.

PLOW HORSES

Plow horses are menu items that are popular but not profitable. They may bring in customers, but selling a lot of them will not improve profitability. Use these management suggestions:

- Consider a selling price increase. One reason for high sales may be the value these items provide. A small price increase may not affect popularity but will increase profitability.
- If portions are large, they could be reduced slightly to decrease food cost and increase the CM.
- If observation of the menu design indicates that they are present in a very high-profile position on the menu, they could be moved to a less obvious spot.
- Combine the item with another menu item that has a much lower food cost to increase profitability. For example, a high-cost twice-baked potato that comes with a menu item might be offered as an *à la carte* item and replaced with a lower-cost potato or starch.
- If the item requires many labor hours to produce, it could be removed from the menu, or ways to reduce labor costs could be considered.

PUZZLES

Puzzles are menu items that are profitable, but not many are sold. Use these management strategies to increase popularity:

- Reduce the price to provide greater customer value.
- Promote the items by moving them to a more visible menu position, using table tents and suggestive selling.
- Rename the item to increase recognition and sales.
- Remove the item if it creates production or service problems.

DOGS

Dogs are menu items that are unprofitable and unpopular. Use these management strategies:

- Remove the item from the menu.
- Raise the selling price or reduce the cost of ingredients to increase profitability.
- Replace with an alternative menu item that supports the brand image. Properly test the item before it is added to the menu.

SUMMARY

1. **Tell why menus are important, explain who should be on the menu planning team, and describe basic menu planning tactics.**

   Menus are important because they attract customers, impact the establishment's financial success, and influence many aspects of daily operations.

   Members of the menu planning team should include the owner-manager, kitchen manager, dining-room manager, purchasing agent, and accountant because persons with these responsibilities will have special views about menu planning concerns.

   There are several important menu planning steps that include considering the customers and other external priorities, thinking about how the menu will impact internal operations, and ensuring that the menu will be in line with the establishment's basic concept. Other menu planning steps require the manager to determine the number of menu item classifications, evaluate potential menu items for each classification, and select specific menu items that will be on the menu.

2. **Describe basic procedures useful to design a menu.**

   There are four common types of menus: à la carte menu, table d'hôte menu, cyclical (cycle) menu, and du jour menu.

   Menus must be planned to be effective customer information and operation sales tools. To do this, it is important to consider where specific menu items will be located, to provide accurate and tempting menu item descriptions, and to incorporate nutrition concerns into the menu. Other tactics include ensuring that menu descriptions are accurate, that the menu appearance and construction meets the establishment's standards, and that space is available for other purposes such as cross-selling opportunities and general property information.

3. **Provide information useful in establishing menu selling prices.**

   The best pricing methods consider the operation's financial goals and use information available in the property's operating budget.

One common pricing method uses the budget's approved food cost percentage to determine the menu item selling price. Three additional menu pricing methods consider the contribution margin (Revenue – Food cost), the ratio between food cost and contribution margin, and the prime cost percentage to develop base selling prices.

A challenge with any objective menu pricing method is to determine the actual food cost for items being processed. This requires knowledge of a menu item's ingredients, current food cost, and yield (number of servings) of the recipe.

4. **Explain procedures to analyze the menu sales mix and implement menu improvements.**

   A menu sales mix analysis is undertaken to determine the popularity and profitability of food items that compete with each other on an establishment's menu. A sales mix analysis involves determining the popularity and profitability of each menu item and then comparing these benchmarks against the average popularity and profitability of a menu item on the menu.

   Menu items can be classified according to popularity and profitability and, when this information is known, menu improvements can be made.

APPLICATION EXERCISES

Exercise 1

Sally has the responsibility to determine the MM percentage for the menu items listed. She must categorize each menu item relative to its MM % category. She has been given the sales records for the last three weeks to make these determinations. Using a chart like this one, calculate the MM % and label the MM % category as either high or low based on your calculation of the menu mix popularity percentage.

Menu Item	**Number Sold**	**MM %**	**MM % Category (High or Low)**
Fried Chicken	375		
Porterhouse Steak	310		
Baked Halibut	110		
Beef Kabobs	70		
Roast Pork	135		
Total			
Menu Mix Popularity % = ________________________			

Exercise 2

Sally has the menu prices of the entrées, the current food cost for each menu item, and the sales records from the last three weeks. Using a chart like this one, find the CM of each menu item and the total CM for the menu items. Also, calculate the average CM and use it to categorize the items.

Menu Item	Number Sold	Selling Price	Item Food Cost	Item CM	Total Item CM	CM Category (High or Low)
Fried Chicken	375	$10.50	$5.15			
Porterhouse Steak	310	16.45	8.50			
Baked Halibut	110	15.95	7.27			
Beef Kabobs	70	9.95	4.16			
Roast Pork	135	13.50	5.10			
Total						
Average CM: ____________						

Exercise 3

Use the information from Exercises 1 and 2 to complete a chart like this one.

Menu Item	MM % Category (High or Low)	CM Category (High or Low)	Menu Item Classification
Fried Chicken			
Porterhouse Steak			
Baked Halibut			
Beef Kabobs			
Roast Pork			

REVIEW YOUR LEARNING

Select the best answer for each question.

1. **An establishment's target market is the group of people who**
   - A. are regular customers of the operation.
   - B. spend a lot of money dining out.
   - C. live close to the establishment.
   - D. enjoy and can afford the menu items.

2. **What is the next menu planning step after menu planning priorities and the establishment's concept have been considered?**
   - A. Determine which items should get prime real estate in the layout.
   - B. Decide on the number of menu item classifications to use.
   - C. Study all menu items that appear on your competition's menu.
   - D. Evaluate trends and favorite items that might interest customers.

3. **What is the purpose of an establishment's brand?**
   - A. Create an image of value in customers' minds.
   - B. Make customers aware of the menu.
   - C. Enable a high profit and investment return.
   - D. Allow the operation to charge higher prices.

4. **Which is the equation for determining contribution margin?**
   - A. Food cost – Revenue
   - B. Food cost + Labor cost
   - C. Revenue – Labor cost
   - D. Revenue – Food cost

5. **Which should be the first group of menu items planned?**
   - A. Appetizers
   - B. Soups
   - C. Salads
   - D. Entrées

6. **What is the name for menus that have different prices for each menu item?**
   - A. À la carte
   - B. Du jour
   - C. Table d'hôte
   - D. Cyclical

7. **Where is prime real estate on a single-sheet menu?**
   - A. Center of menu's lower half
   - B. Center of menu's upper half
   - C. Top right corner of menu
   - D. Bottom right corner of menu

8. **What is the approximate selling price of a menu item if it has a food cost of $3.75 and a 32% food cost is desired?**
   - A. $11.75
   - B. $12.00
   - C. $12.25
   - D. $12.50

9. **What is the average CM per customer if there are 78,000 customers, profit is $39,000, and nonfood costs are $390,000?**
   - A. $4.50
   - B. $5.00
   - C. $5.50
   - D. $6.00

10. **A plow horse is a menu item that is**
    - A. unpopular and unprofitable.
    - B. popular and profitable.
    - C. popular but unprofitable.
    - D. profitable but unpopular.

FIELD PROJECT

1. Conduct a brief interview with the manager of a table-service establishment and ask him or her what are the three most important things that should be considered when a menu is designed.

2. Ask one or two relatives or friends the following questions:

   A. What are three things you want to see when you look at a menu in a table-service establishment?

   B. What are three things that disappoint you when you see them on a menu at a table-service establishment?

   C. What are three menu design features that would be of interest to you when you review a menu from a table-service establishment?

3. Consider the information you learned from the previous interviews and state three principles that the menu planning team should implement when a menu for a table-service operation is designed.

   Menu planning and design principle 1:

   ____

   ____

   ____

   Menu planning and design principle 2:

   ____

   ____

   ____

   Menu planning and design principle 3:

   ____

   ____

   ____

2 Standardized Recipes Are Critical

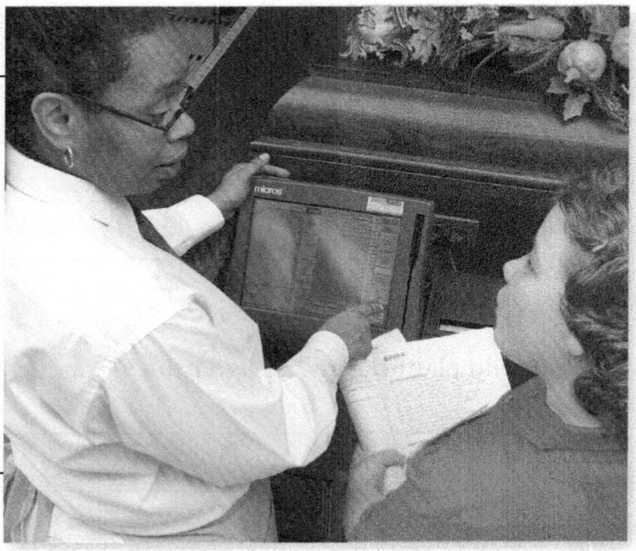

INSIDE THIS CHAPTER

- Importance of Standardized Recipes
- Developing Standardized Recipes
- Standardizing Recipes for an Operation
- Calculating Recipe Costs
- Completing Recipe Development Requirements
- Technology and Standardized Recipes

CHAPTER LEARNING OBJECTIVES

After completing this chapter, you should be able to:

- Explain the importance of standardized recipes.
- List the procedures for developing standardized recipes.
- Describe how to standardize recipes for a specific operation.
- Calculate recipe ingredient costs, standardized recipe costs, and menu (plate) costs.
- Explain how to ensure food safety requirements are addressed in standardized recipes, and explain procedures helpful when implementing and consistently using standardized recipes.
- Describe how technology impacts the use of standardized recipes.

KEY TERMS

accompanament, p. 56

as purchased (AP), p. 48

batch cooking, p. 45

blind tasting, p. 44

calibrate, p. 60

coaching, p. 58

comfort food, p. 42

critical control point (CCP), p. 58

edible food yield, p. 51

edible portion (EP), p. 48

Hazard Analysis Critical Control Point (HACCP) system, p. 58

ingredient file, p. 62

menu costing, p. 56

menu item file, p. 63

point-of-sale (POS) system, p. 63

productivity, p. 41

purchase unit (PU), p. 50

recipe conversion factor (RCF), p. 46

recipe evaluation, p. 43

recipe management software, p. 62

standardized recipe file, p. 62

tare allowance, p. 61

yield, p. 38

CASE STUDY

"After all these years, I guess our boss decided he doesn't think we know very much!" said Chelsea during a coffee break conversation with Chester.

"Yes, I have been thinking the same thing," he replied. "Our boss's idea about writing down our recipes and using the same recipe every time doesn't make much sense to me. We all use the same basic ingredients and just change it a little bit to add our special touch. What's wrong with that?"

"I think you're right, Chester, and I'm wondering about something else. Maybe our boss wants these recipes so he can get rid of us. Then he can replace us with people who know a lot less than we do, and he can pay them a lot less money."

"I've thought about that too," said Chester, "and I wish I knew what is going on!"

1. What is the main problem causing concern for Chelsea and Chester?
2. What should the manager do to address the problem?

THINK ABOUT IT . . .

Many managers recognize the importance of using standardized recipes to prepare food items. However, they may not use them for preparing alcoholic beverages.

Do you think standardized recipes are needed for alcoholic beverages? Why or why not?

IMPORTANCE OF STANDARDIZED RECIPES

Recall from chapter 1 that a standardized recipe is the set of instructions to produce a food or beverage item. When followed, there is better assurance that a product's quality and quantity standards will be met.

Every restaurant and foodservice operation needs standardized recipes, and there are really no exceptions. Standardized recipes indicate what ingredients are needed and in what amount to produce required items. They also give the preparation procedures and the **yield**, which is the number of servings and serving size. Standardized recipes also provide any other information required to prepare the item. The use of standardized recipes helps ensure that products will meet cost and customer expectations.

Managers want to offer menu items customers want, and they consider many factors as menus are planned. It then becomes important to ensure that items are consistently produced according to quality and quantity standards. This can be done only when standardized recipes are used.

Benefits of Standardized Recipes

The consistency offered by a standardized recipe benefits customers and the operation. Each time an item is ordered, it will taste and look the same. The size of the serving will be the same, and the same ingredients in the same amounts will be used. The customers will receive consistent value because they will pay the same amount every time they order an item. Standardized recipes ensure that the product will be of the same quality and quantity each time (*Exhibit 2.1*).

Exhibit 2.1

Manager's Memo

In some operations, a chef-owner closely supervises the cooks, and recipes may be informal or not even in writing. However, a specific set of procedures is still used to meet quality requirements. More commonly, especially in chain restaurants and high-volume operations, production staff follow standardized recipe procedures with or without referring to a copy of the recipe. Note: *A chain restaurant is one unit in a multiunit restaurant organization.*

In this chapter, a standardized recipe will refer to the usually written but always followed procedures for how a menu item should be produced.

A standardized recipe also helps ensure consistency for the operation. First, recall from chapter 1 that a menu item's selling price is often based, at least in part, on ingredient costs. A standardized recipe ensures that the same ingredients and amounts of each ingredient are used every time the product is prepared. Then food costs and selling prices can be determined.

Properly developed standardized recipes give all the information needed for food-production employees to prepare menu items, including the following:

- **Ingredient details:** These include grades and brands of products needed, precise descriptions, and so on.
- **Correct weights and measures of the ingredients**
- **Equipment and tools needed:** Everything from pots to utensils to large kitchen equipment must be specified; all information that is particularly important to kitchen workers.
- **Volume to be produced:** Volume depends partly on the portion size, or the size of an item's individual serving, such as "four ounces of chicken."
- **Time required to prepare the item**
- **Storage and preparation information:** This includes information on thawing, what types of cuts need to be made, and so forth.
- **Cooking method to be used:** Examples include blanching, sautéing, deep-frying, and other methods.

Labor cost control, especially employee scheduling, is easier with standardized recipes. Standardized recipes ensure that it will take about the same time to prepare items regardless of who makes them. Standardized recipes also make it easier to train new production employees. New employees can be given copies of recipes.

Surprisingly, standardized recipes are not used in some properties. Several reasons are given:

- Cooking professionals do not need them.
- There is not enough time to read them.
- "Secret" recipes will be stolen if they are written down.
- It is time-consuming to keep them up to date.
- Reading levels of some production staff are not adequate.

Only an inability to read recipes has some validity. Even when there are language barriers, frequently used recipes can be translated.

Accuracy in Menu Disclosures

Many customers have concerns about how menu items are produced and the ingredients they contain. There has been an increase in legislation requiring managers to disclose information about some ingredients used in food preparation. For example, some communities have banned the use of trans fats in cooking. Standardized recipes help production staff remain in compliance with these laws. Only the ingredients required by the recipes will be used.

Manager's Memo

Some customers have dietary and health concerns that require food-production staff to know the exact ingredients and amounts of each used to produce items ordered. For example, some customers may be on diets that severely restrict their sodium (salt) intake, and they may ask servers if salt is used in a specific menu item. Cooks can answer this question accurately when standardized recipes are followed.

Sometimes, however, there may be very small amounts of ingredients such as salt in processed food in a menu item. If the amount of salt in the processed food is unknown, the best response will be to suggest that the customer select another menu item.

THINK ABOUT IT . . .

Assume your family operates a small neighborhood establishment. You cook, your spouse manages the dining room, and your children serve the meals. Would you use standardized recipes? Why or why not?

DEVELOPING STANDARDIZED RECIPES

Standardized recipes are clearly important. Managers generate these recipes in a variety of ways. Managers can develop recipes for items that are currently being produced in the operation. In addition, recipes from external sources can be modified as necessary when new menu items are developed.

THINK ABOUT IT . . .

Some cooks may resist standardized recipes. Some want to defend how things have always been done. Others may have a concern for their job.

What would you say to cooks who express these concerns?

Recipes for Current Menu Items

Imagine that a manager decides to develop standardized recipes for menu items currently being produced without them. An important "first step" is to tell food-production personnel why the recipes are needed. Employees can also be asked to assist in standardized recipe development activities. Then there will likely be less resistance to the recipe's use. The recipes will have been developed by the food-production team and not just by the chef or manager.

Exhibit 2.2

STANDARDIZED RECIPE DEVELOPMENT: CURRENT MENU ITEMS

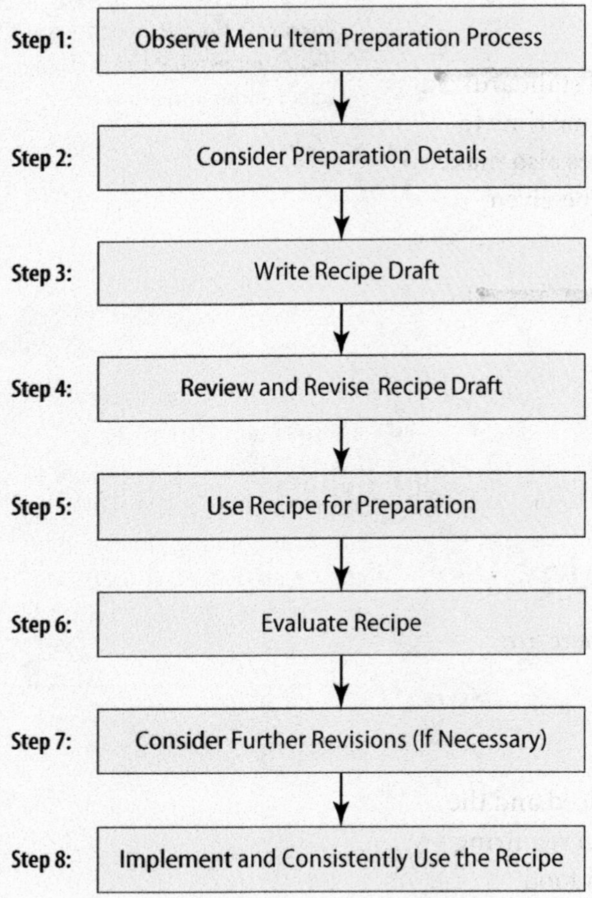

Several steps for developing a standardized recipe for a menu item currently being produced are illustrated in *Exhibit 2.2*. The steps are fairly simple because the menu item is already being prepared in the correct way.

In reviewing *Exhibit 2.2*, recall that these steps apply to an operation that is already using effective procedures to meet production standards. The purpose of the recipe development process for these properties is to commit current production procedures to a written or computerized form. Managers commit to standardized recipes to take advantage of costing, scheduling, training, and other benefits.

Step 1: Observe Menu Item Preparation Process

Food-production employees can be observed as they prepare the menu item. What ingredients are used and in what quantities? How and when are the ingredients weighed and measured? What small and large equipment is needed and when? How is the item portioned after preparation? When different cooks produce the same item, production techniques used by each employee should be observed.

Step 2: Consider Preparation Details

Can some preparation tasks be combined? If utensils needed for several recipe steps are all stored in the same general area, include a procedure suggesting, "Obtain all necessary utensils [state what they are]." This can increase productivity by saving the time needed for repeat trips to the utensil storage area. **Productivity** refers to the quality and quantity of output compared to the amount of input such as labor hours needed to generate it.

Step 3: Write Recipe Draft

A first draft of the recipe that incorporates observations from step 1 and details from step 2 now becomes important. A standardized format should be used; an example is shown in *Exhibit 2.11* later in this chapter.

Step 4: Review and Revise Recipe Draft

The recipe draft should be reviewed by the production personnel who prepare the item and revisions should be made as necessary.

Step 5: Use Recipe for Preparation

The recipe draft should be carefully followed to prepare the menu item.

Step 6: Evaluate Recipe

The food item produced in step 5 should be carefully evaluated to determine if quality standards have been attained.

Step 7: Consider Further Revisions (If Necessary)

If the recipe does not yield a product meeting quality standards, further revisions to the recipe are necessary.

Step 8: Implement and Consistently Use the Recipe

After the recipe is developed with agreement about desired quality, it should be used each time the product is prepared.

Recipes for New Menu Items

Consider a situation in which menu planners have revised the menu. They have decided to add a new item. How does the chef or manager develop or locate a recipe for the new item? Procedures to develop recipes for new menu items are somewhat different from those used to develop recipes for existing items. A basic approach is to identify possible recipes that can be used to produce the new menu item, with changes as necessary.

The manager or chef may have already reviewed possible recipes as part of the menu revision process. Those sample recipes provide a starting point for further review and development.

Exhibit 2.3

STANDARDIZED RECIPE DEVELOPMENT: NEW MENU ITEMS

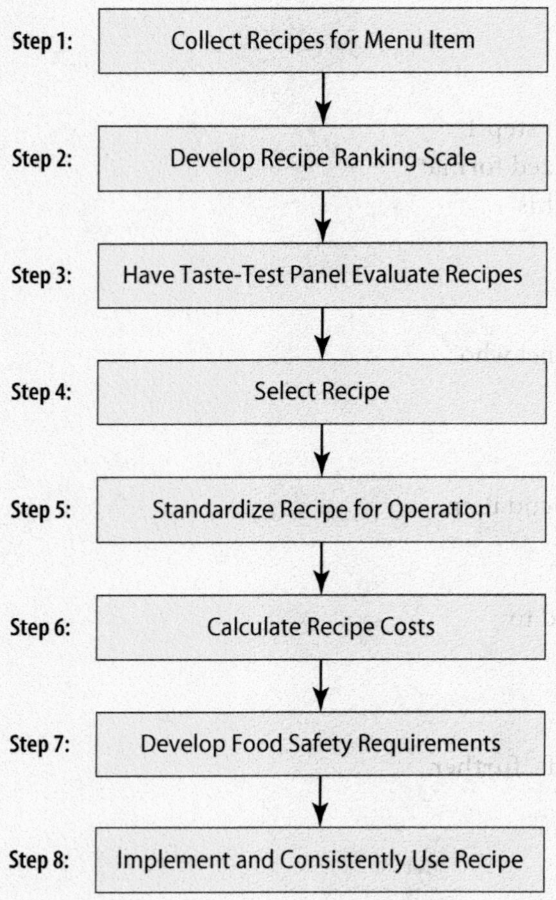

Exhibit 2.3 previews steps useful in developing standardized recipes for a new menu item.

Each of these steps will be reviewed in detail.

Step 1: Collect Recipes for Menu Item

Menu planners can find recipes with yields of 25, 50, and more in many cookbooks and on the Internet. Many chefs have a network of persons in other properties who share recipes. Those belonging to local restaurant or chefs' associations will likely know many persons who can be asked about recipes.

Numerous recipes for almost any menu item can be easily located. An early evaluation step is to match recipes with the operation's concept. For example, a family restaurant featuring comfort food will not want a recipe using unusual ingredients and preparation methods. **Comfort food** refers to familiar menu items prepared as a customer might do at home.

External recipes should be reviewed for potential use. The trained and experienced eye of the food-production manager can help assess whether potential recipes should be considered.

Recipes from external sources should be thoroughly tested before they are implemented. If the yield is relatively small, the menu item can be prepared by strictly following all procedures in the recipe. Careful evaluation, changes, and further testing may be in order. If the recipe yields a large number of portions it can be adjusted downward before testing. The kitchen manager can follow steps 5 to 7 in *Exhibit 2.2* to test recipes.

Step 2: Develop Recipe Ranking Scale

Assume that several recipes are of interest to the food-production team. Which should be selected? As demonstrated in the next step, one or more taste-test panels can evaluate food items produced by the recipes. What factors should the taste-test panels consider?

The answer to this question should be incorporated into a recipe evaluation form such as the one shown in *Exhibit 2.4*.

Exhibit 2.4 shows common factors used to evaluate recipes. A five-point scale from *poor* to *excellent* is used for each factor, and taste-test panel members can provide additional comments.

Exhibit 2.4

STANDARDIZED RECIPE EVALUATION FORM

Standardized Recipe Name: ______________ Menu Category: ______________

Evaluation Date(s): ______________ Recipe No.: ______________

Instructions: Check the box that best represents your analysis of each factor. Note any additional thoughts in the comments box.

EVALUATION FACTOR	YOUR ANALYSIS Poor Excellent	COMMENTS
Serving Size	☐ ☐ ☐ ☐ ☐	
Color	☐ ☐ ☐ ☐ ☐	
Texture	☐ ☐ ☐ ☐ ☐	
Taste	☐ ☐ ☐ ☐ ☐	
Aroma	☐ ☐ ☐ ☐ ☐	
General Appearance	☐ ☐ ☐ ☐ ☐	
Ingredients	☐ ☐ ☐ ☐ ☐	
Compatibility with concept	☐ ☐ ☐ ☐ ☐	
Garnish, appropriateness of	☐ ☐ ☐ ☐ ☐	
Other: ______________	☐ ☐ ☐ ☐ ☐	
Other: ______________	☐ ☐ ☐ ☐ ☐	

Should we use this recipe? ☐ Yes ☐ No

Comments:

Name of Evaluator: ______________________________

Step 3: Have Taste-Test Panel Evaluate Recipes

Recipe evaluation is a formal process in which members of the taste-test panel assess whether a recipe produces an acceptable product. Who should serve on the taste-test panel? The restaurant and food-production managers, assistant managers, cooks, and food and beverage servers are possible members. Guests including the operation's "regulars" are also great choices for taste-test panels.

Hopefully, a representative group from the operation's target market will serve on a taste-test panel. Guests often do so in exchange for complimentary meals or some other similar compensation. Panels composed of non-employees may be scheduled at times when employees are not participating. Then they can be more candid in any discussions after the evaluation form is completed.

The use of customers on a taste-test panel has an additional benefit. It provides an opportunity to show the establishment's concern for customer opinions and enabling customers to learn more about how it operates.

Dining-room tables should be set with the place settings used for regular meal service (*Exhibit 2.5*). Test recipes should be prepared, when possible, with the actual equipment used in normal operations. This may not be possible when normal production will be for a large number of servings and taste-test panel portions will be only a few.

Exhibit 2.5

Blind tasting of food items ensures that raters do not know the recipe's ingredients or preparation methods when they sample the items. This can help reduce preconceived opinions that arise when panel members have knowledge about what they are sampling.

Step 4: Select Recipe

Should the recipe selected as best by the taste-test panel be added to the menu? While the recipe used must produce products judged favorably by customers, specific operating concerns are also factors. Managers must determine the answers to questions such as these:

- Do we have the equipment capacity to produce the item?
- Do food-production employees have the necessary knowledge and skills, or can they acquire them?
- Would the operation produce the quantity needed at one time or use a batch cooking method? In **batch cooking**, items needed in large quantities are produced in small volumes (batches) to maximize food quality by reducing holding times until service.
- Are there any special serving problems?
- Does the establishment have the proper serviceware?
- Can the establishment safely reuse leftovers, if applicable?

Another concern relates to product cost, which will drive the selling price and the customers' perceptions of value. While detailed costing may not be necessary at this point, a general overview is needed.

Imagine that the new item will be an entrée: one-half baked chicken with a special sauce.

- Purchase size = 2.5 pound chicken, chilled on ice
- Serving size = 1/2 of whole chicken

2.50 lb	÷	**2**	=	**1.25 lb**
Weight of chicken		**No. of servings**		**Weight per serving**

1.25	×	**$2.80**	=	**$3.50**
Weight per serving		**Price per lb**		**Cost per serving**

The food-production manager estimates the food cost for the entrée and items that accompany the entrée:

Entrée:	$3.50
Sauce:	0.85
Vegetable choice:	1.05
Potato choice:	0.90
Bread/butter:	0.22
Total estimated food cost	**$6.52**

While these calculations provide only an estimate of food cost, this benchmark can be used to establish a base selling price (see chapter 1). The manager and his or her team should carefully consider whether the price that might be charged for the new chicken entrée would be a value to customers.

A decision to select one recipe from several available options is made based on taste-test panel evaluations and operating concerns. After the recipe is selected, it must be standardized for the specific operation.

These are the first four steps in the eight-step process for developing standardized recipes for new menu items. The last four steps require knowledge about numerous details, and they are discussed in the following sections.

STANDARDIZING RECIPES FOR AN OPERATION

Step 5 in the process of developing a recipe for a new menu item is to standardize the recipe for the specific operation. Two important concerns are adjusting the recipe for the property and carefully considering recipe measurements.

Recipe Adjustments

Recipes with desired yields are sometimes selected. However, adjustments must be made frequently:

- To increase or decrease the number of servings
- To increase or decrease serving sizes
- To increase or decrease both number of servings and serving sizes

Sometimes, a recipe has the correct serving size, such as 4 ounces or 1/2 cup, but it yields more or fewer servings than desired. *Exhibit 2.6* shows how to make this adjustment.

Exhibit 2.6

ADJUSTING NUMBER OF SERVINGS

The current recipe yields 40 servings (3 oz each) but 70 servings (3 oz each) are needed. What quantity of each ingredient is needed in the converted recipe?

1. Determine the **recipe conversion factor (RCF)**: a factor (number) used to adjust ingredients in a recipe when the number of servings and/or serving size for a current recipe differs from the number desired. The RCF is calculated by dividing the desired number of servings by the current number of servings.

70	÷	40	=	1.75
Desired servings (3 oz)		Current servings (3 oz)		RCF

2. Multiply the amount of each ingredient in the current recipe by the RCF.

8 oz flour	×	1.75	=	14 oz flour
Current recipe		RCF		Converted recipe

If the serving size does not change, the RCF will always be *less than 1.0* if *fewer* servings are desired and *greater than 1.0* if *more* servings are needed.

Sometimes the recipe selected has the right number of servings, but different serving sizes are desired. The current recipe may yield 50 5-ounce servings, but 50 4-ounce servings are desired. *Exhibit 2.7* shows how to make the adjustment.

Exhibit 2.7

ADJUSTING SERVING SIZE

The current recipe yields 50 servings of 5 oz each, but 50 servings of 4 oz each are desired.

1. Determine RCF by dividing the desired yield (Desired servings × Serving size) by the current yield (Current servings × Serving size).

$$\frac{\text{Desired servings} \times \text{Serving size}}{\text{Current servings} \times \text{Serving size}} = \frac{50 \times 4\text{ oz}}{50 \times 5\text{ oz}} = \frac{200\text{ oz}}{250\text{ oz}} = 0.80$$

2. Multiply the amount of each ingredient in the current recipe by the RCF.

10 oz chopped onion	×	0.80	=	8 oz chopped onion
Current recipe		RCF		Converted recipe

If the number of servings does not change, the RCF will always be *less than 1.0* if a *smaller* serving size is desired and *greater than 1.0* if a *larger* serving size is needed.

Sometimes both servings and serving sizes need to be adjusted. This process is shown in *Exhibit 2.8*.

Exhibit 2.8

ADJUSTING NUMBER OF SERVINGS AND SERVING SIZES

The current recipe yields 60 servings of ½ cup each, but 80 servings of ¾ cup each are desired.

1. Determine the RCF:

$$\frac{\text{Desired servings} \times \text{Serving size}}{\text{Current servings} \times \text{Serving size}} = \frac{80 \times \tfrac{3}{4}\text{ c}}{60 \times \tfrac{1}{2}\text{ c}} = \frac{60\text{ c}}{30\text{ c}} = 2.00$$

2. Multiply the amount of each ingredient in the current recipe by the RCF.

3 oz flour	×	2.00	=	6 oz flour
Current recipe		RCF		Converted recipe

The basic arithmetic procedures just described are simple. However, food-production staff members work in a busy environment, so mistakes are easily made. It is always important to double-check calculations and review recipe adjustments made by food-preparation assistants before a recipe is used.

Recipe Measurements

Recipe ingredients must always be carefully weighed and measured. It is important to express amounts in the easiest way for cooks to weigh or measure them. *Exhibit 2.9* on the following page shows common U.S. measurements and weights.

MANAGER'S MATH

Almost every country in the world except the United States uses the metric measurement system. The metric system uses decimals of three basic units:

- Meter: length; about 3.2 feet
- Gram: weight; about the weight of a paper clip
- Liter: volume; about 1.05 quarts

Very large and very small units are expressed in multiples of ten of the base unit.

Greater Than Base Unit	**Base Unit**	**Less Than Base Unit**
Kilo 1,000 times	Meter (m): length	Deci: 0.1 times
Hecto 100 times	Gram (g): weight	Centi: 0.01 times
Deka 10 times	Liter (L): volume	Milli: 0.001 times

The Fahrenheit temperature scale used in the United States is based on 32°F for the freezing point of water and 212°F for its boiling point. The Celsius temperature scale is based on 0°C for the freezing point of water and 100°C for its boiling point.

Exhibit 2.10

U.S. MEASUREMENTS AND METRIC EQUIVALENTS

U.S.	**Metric**
Volume	
Gallon	3.79 liters (L)
Quart	0.95 L
Pint	473.2 milliliters (ml)
Cup	236.6 ml
Tablespoon	14.8 ml
Teaspoon	4.9 ml
Weight	
Pound	454 grams (g)
¾ pound (12 oz)	340 g
½ pound (8 oz)	227 g
¼ pound (4 oz)	113 g
1 ounce	28 g

Exhibit 2.9

COMMON U.S. MEASUREMENTS AND WEIGHTS

Part I: Volume (Gallon to Teaspoon)				
1 gallon	=	4 quarts	=	128 fluid ounces
1 quart	=	2 pints	=	32 fluid ounces
1 pint	=	2 cups	=	16 fluid ounces
1 cup	=	16 tablespoons	=	8 fluid ounces
1 tablespoon	=	3 teaspoons	=	½ fluid ounce
Part II: Volume (Cup to ½ Tablespoon)				
1 cup	=	16 tablespoons		
¾ cup	=	12 tablespoons		
⅔ cup	=	10 tablespoons + 2 teaspoons		
½ cup	=	8 tablespoons		
⅓ cup	=	5 tablespoons + 1 teaspoon		
¼ cup	=	4 tablespoons		
⅛ cup	=	2 tablespoons		
1 tablespoon	=	3 teaspoons		
½ tablespoon	=	1½ teaspoons		
Part III: Weight (Pound to Ounce)				
1 pound	=	16 ounces		
¾ pound	=	12 ounces		
½ pound	=	8 ounces		
¼ pound	=	4 ounces		

Managers may select excellent recipes that use the metric system. They will need to convert ingredient amounts to U.S. weights and measures. *Exhibit 2.10* shows some U.S. and metric equivalents.

The Internet also has tools for converting U.S. measurements and temperatures to the metric and Celsius system, or the reverse.

A standardized recipe's yield is the quantity produced when the recipe is followed. Yields can be expressed in several ways:

- By volume: 2 gallons of beef stew
- By weight: 16 pounds of bread dough
- By number of servings: 25 ½-cup servings

Serving size is important in determining yield. A hamburger patty may weigh 6 ounces **as purchased (AP)** and before being cooked. It may have an **edible portion (EP)** weight after cooking of 5.4 ounces.

Kitchen staff must control the quantity of ingredients to produce the expected yield. Careful measuring, weighing, and counting are important when recipes are prepared. Appropriate tools must be used to ensure accuracy. If a recipe requires 1 cup of milk and 6 ounces of flour, the required liquid measuring tool and scale must be available.

Format for Standardized Recipes

Exhibit 2.11 shows one format for a standardized recipe. Food-production managers often use the same format for all recipes. When a new recipe is developed, that standard design should be used.

Exhibit 2.11

SAMPLE STANDARDIZED RECIPE: CHICKEN TETRAZZINI

UTENSILS NEEDED: Stock pot, cutting board, French knife, wire whip, mixing bowl, gallon, quart, and cup measures, measuring spoons, plastic gloves, clean foodservice cloths, 2 steam table pans (12″ × 20″ × 2″)

YIELD: 48 servings
OVEN TEMP: 350°F (176.6°C)
BAKING TIME: 30 minutes
SERVING SIZE: 1/24 pan
SERVING TOOL: Spatula or spoon

Ingredients	Quantity/ Volume	Procedure
Spaghetti	6 lb	1. Cook spaghetti in salted water. Rinse and drain. Do not overcook.
Margarine	2 lb	
Celery, cut fine	2 qt	2. Cook onions and celery in margarine until transparent.
Onions, cut fine	2 qt	
Flour, pastry	1 lb 4 oz	3. Make roux by adding flour, salt, and pepper. Cook 5 minutes.
Salt	2.5 oz	
Pepper, black	1 tsp	4. Add chicken (turkey) stock and cook until thick, stirring as necessary.
Chicken (turkey) stock	2 gal 2 c	
Chicken (turkey), cooked and cubed	12 lb 8 oz	5. Add cubed chicken (turkey) and mushrooms; mix.
Mushrooms, fresh/chopped	2 c	6. Add spaghetti; mix well.
Green pepper, chopped	3 c	7. Add green pepper just before panning.
		8. Scale 12 lb. into each 12″ × 20″ × 2″ pan.
Total Weight	48 lb	
		9. Mix topping. Top each pan with 1 qt. topping.
Topping		
Bread crumbs, fine	2 qt	10. Bake at 350°F (176.6°C) for 30 minutes.
Sharp cheddar cheese, grated	2 qt	11. Serving: Divide into servings by cutting pan contents 6 (length) × 4 (width).

Example:

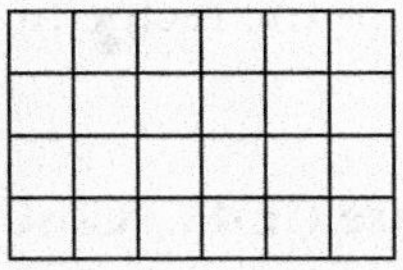

Holding: Hold prepared product at 140°F until service.

SPECIAL INSTRUCTIONS:
Do not overcook spaghetti. See recipe for chicken stock if none is available.

The recipe in *Exhibit 2.11* indicates all necessary ingredients and quantities, preparation procedures, and utensils and other equipment. It also tells the baking time and temperature and the portioning (serving) instructions:

- The yield in number of servings (48) and serving size (1/24 of a 12″ × 20″ × 2″ steam table pan).
- The method of portioning, which is illustrated for clarity.

CALCULATING RECIPE COSTS

Step 6 in the development of standardized recipes involves determining cost. Recall from chapter 1 that recipe costing (also called pre-costing) considers the cost to produce all or a single serving of a recipe. Costing works by calculating ingredient costs and the number of servings the recipe yields.

Accurate recipe costing is important:

- If costs to produce menu items are considered when menus are planned, managers can determine if items of desired quality can be produced and sold at a price providing value to the customers.
- Managers can assess costs of buffet and salad bar items. If costs for some ingredients are excessive, the items can be replaced or prices can be reevaluated.
- As noted in chapter 1, objective selling prices consider the food costs of the items being sold.

Costing Recipe Ingredients

The task of costing ingredients is easy when the purchase unit is the same unit used when the item is included in a recipe. The term **purchase unit (PU)** refers to the weight, volume, or container size in which a product can be purchased. For example, salad oil can be purchased in gallon jars, and ground beef is purchased by the pound.

Fluid whole milk is an item that often has the same purchase and usage unit:

Purchase Unit	Cost per Unit	Quantity in Recipe	Ingredient Cost
Gallon	$4.75	2 gal	$9.50*

***$4.75 per gal × 2 gal = $9.50**

Recipe costing is more difficult when an ingredient is purchased in one unit and used in the recipe in a different unit. Consider fluid whole milk in this example:

Purchase Unit	Cost per Unit	Quantity in Recipe	Ingredient Cost
Gallon	$4.75	2 qt	$2.38*

***$4.75 per gal ÷ 4 qt = $1.19; $1.19 per qt × 2 qt = $2.38**

The costing process becomes even more challenging when an ingredient is purchased in a weight unit and used in a volume unit. For example, assume celery costs $1.69 per pound, and 1 cup of chopped celery is needed in the recipe. Two types of information are needed to cost this ingredient:

- How to convert pounds (weight) to cups (volume).
- How much celery is lost in cleaning and removing leaves and ends. One pound (16 ounces) of celery may yield only 14 ounces that can be chopped for the recipe.

Fortunately, edible food yield guides are available. An **edible food yield** is the usable amount of a food ingredient that can be prepared from a given PU of the ingredient. For example, 1 pound (16 ounces) of fresh mushrooms (*Exhibit 2.12*) yields 6 cups of cleaned sliced mushrooms or 5 cups of cleaned chopped mushrooms.

Use of an edible food guide eliminates the time needed to conduct tests to determine the quantity of prepared ingredient for every 1 pound of ingredient purchased.

Exhibit 2.13 shows how to cost the recipe in *Exhibit 2.11.* When reviewing this example, remember the following abbreviations: lb = pound; qt = quart; oz = ounce; c = cup; gal = gallon; tsp = teaspoon.

Exhibit 2.12

THINK ABOUT IT . . .

To view volume and weight conversions for commonly used food ingredients, type "food yields" into a search engine. How would you use edible food yield information if you were costing a recipe?

Exhibit 2.13

RECIPE COSTING WORKSHEET

Recipe: Chicken Tetrazzini

Yield: 48 servings Serving size: 1/24 pan (12″ × 20″ × 2″)

(1) Ingredient	(2) Amount	(3) Purchase Unit	(4) Cost per PU	(5) No. of PUs	(6) Ingredient Cost
(A) Spaghetti	6 lb	(lb)	$1.03	6	$ 6.18
(B) Margarine	2 lb	(lb)	0.89	2	1.78
(C) Celery	2 qt	Bunch	0.99	1.5	1.49
(D) Onions	2 qt	(lb)	1.69	2.2	3.72
(E) Flour	1 lb 4 oz	(lb)	2.10	1.25	2.63
(F) Salt	2.5 oz	(lb)	0.88	0.16	0.15
(G) Pepper	1 tsp	(lb)	—	—	—
(H) Chicken stock	2 gal 2 c	—	—	—	—
(I) Chicken	12 lb 8 oz	(lb)	2.35	26	61.10
(J) Mushrooms	2 c	(lb)	4.95	0.33	1.63
(K) Green pepper	3 c	(lb)	3.05	1.2	3.66
(L) Bread crumbs	2 qt	(lb)	1.90	1.75	3.33
(M) Sharp cheddar cheese, shredded	2 qt	(lb)	5.25	2.0	10.50
			Total Cost:		**$96.17**
			Cost per Serving:		**$2.00**

The costing for each ingredient will be reviewed in detail. Remember that the purpose of the costing process is to determine the cost of each ingredient used and that information about edible food yields is needed to cost some ingredients.

- **Item A (spaghetti):** 6 pounds are needed (column 2), the PU is pounds (column 3), and the cost per PU is $1.03 (column 4). No unit conversion is necessary; multiplying the amount by the unit cost gives an ingredient cost (column 6) of $6.18.

6 lb	×	**$1.03**	=	**$6.18**
Col. 2		**Col. 4**		**Col. 6**

- **Item B (margarine):** 2 pounds are needed, the PU is pounds, and the cost per PU is $0.89.

2 lb	×	**$0.89**	=	**$1.78**
Amount needed		**Cost per bunch**		**Margarine cost in recipe**

- **Item C (celery):** 2 quarts are needed, the PU is bunches, and the cost per PU is $0.99. To cost this item, the PU (bunches) must be converted to the amount needed in volume units (quarts).

  A table of edible food yields indicates that 1 bunch of celery weighs about 2 pounds (32 ounces AP weight) and yields 22 ounces after cleaning and trimming. One cup of cleaned and trimmed celery weighs 4 ounces. Therefore, the volume measure is 5.5 cleaned and trimmed cups per bunch of celery:

22 oz	÷	**4 oz**	=	**5.5 c per bunch**
Yield per bunch		**Weight per cup**		**Yield in volume unit**

  The recipe requires 2 quarts (column 2), which is eight cups:

  2 qt × 4 c per qt = 8 c

  Now that the unit is the same, dividing the amount needed by the yield per bunch gives the amount that will be used in the recipe, 1.5 bunches (AP):

  8 c needed ÷ 5.5 c per bunch = 1.5 bunches (rounded)

  Multiplying this amount by the unit cost gives the ingredient cost:

1.5 bunches	×	**$0.99**	=	**$1.49 (rounded)**
Amount needed		**Cost per purchase unit**		**Celery cost in recipe**

- **Item D (onions):** 2 quarts are needed, the PU is pounds, and the cost per PU is $1.69.

  The onions provide another example of the need to convert the PU (pounds) to the amount needed (quarts). A table of edible food yields indicates that 1 pound (16 ounces) of onions yields 14.5 ounces after they are cleaned and trimmed. One cup of trimmed and cleaned onions weighs 4 ounces. Therefore, 1 pound of onions (AP) yields 3.6 cups of trimmed and cleaned (EP) onions:

14.5 oz	÷	4 oz	=	3.6 c (rounded)
Yield per pound		**Weight per cup**		**Yield in volume unit**

The recipe requires 2 quarts (8 cups) of onions, which converts to 2.2 pounds (AP):

8 c ÷ 3.6 c per lb = 2.2 lb

Now the ingredient cost can be calculated:

2.2 lb	×	$1.69	=	$3.72 (rounded)
Amount needed		**Cost per purchase unit**		**Onion cost in recipe**

- **Item E (flour):** 1 pound 4 ounces are needed, or 1.25 pounds, and the cost per pound is $2.10.

1.25 lb	×	$2.10	=	$2.63
Amount needed		**Cost per unit**		**Ingredient cost**

- **Item F (salt):** 2.5 ounces are needed; the PU is pounds, and the cost per PU is $0.88.

$0.88	÷	16 oz per lb	=	$0.06
Cost per pound				**Cost per ounce**

$0.06	×	2.5 oz	=	$0.15
Cost per ounce		**Amount needed**		**Ingredient cost**

- **Item G (pepper):** The amount of pepper (1 teaspoon) is judged to be insignificant; no cost is calculated for this ingredient.
- **Item H (chicken stock):** While a large amount of chicken stock is needed (2 gallons + 2 cups), there is no direct cost for this ingredient because it is made in-house from chicken bones and vegetable trimmings.
- **Item I (chicken):** 12 pounds 8 ounces are needed, the PU is pounds, and the cost per PU is $2.35. To cost this ingredient, the EP yield of 1 pound of raw chicken must be converted to its AP weight.

  A table of edible food yields indicates that 1 pound of chicken (whole bird, large fryer) has a 48 percent yield: for every pound purchased, only 48 percent will be left after cooking and bone and fat removal.

  The recipe requires 200 ounces of edible chicken:

  12 lb × 16 oz per lb + 8 oz = 200 oz

Manager's Memo

Does the $0.15 cost for salt in a recipe yielding 48 servings really affect the recipe cost? Managers must answer these types of questions as they cost recipes.

Some say, "It takes only a few seconds, and a lot of small costs add up to a large cost." Others say, "The time saved can be used to control costs in other areas." Many managers do not include small costs. They decide what "small" means. Instead, they add an amount such as a few dollars to the total ingredient expense in the recipe costing process.

It is always important for managers to do what they think is most practical and cost- effective for their operation without affecting quality standards or customer concerns.

First it is necessary to determine what AP weight will yield this amount. Dividing the EP needed by the yield percentage shows that 417 ounces, or 26 pounds, of chicken must be purchased:

200 oz	÷	0.48	=	417 oz
EP		Yield		AP

Divide the required number of ounces by 16 ounces per pound to determine how many pounds of chicken must be purchased for the recipe:

417 oz	÷	16 oz per lb	=	26 lb
				Purchase amount

The ingredient cost is calculated using this AP weight:

26 lb	×	$2.35	=	$61.10
Purchase amount		Cost per pound		Ingredient cost

- **Item J (mushrooms):** 2 cups are needed, the PU is pounds, and the cost per PU is $4.95. To cost the mushrooms, the same approach must be used as for the celery and onions. A table of edible food yields indicates that 1 pound of mushrooms yields 6 cleaned cups. Therefore, the recipe requires 0.33 pound:

2	÷	6 c per lb	=	0.33 lb
Amount needed		Yield per pound		Amound required

With the unit converted, multiply by the unit cost to get the ingredient cost:

$4.95	×	0.33	=	$1.63
Cost per pound		Pounds needed		Ingredient cost

- **Item K (green pepper):** 3 cups are needed, the PU is pounds, and the cost per PU is $3.05. A table of edible food yields indicates that 1 pound (16 ounces) AP yields 81.3 percent (13 ounces) EP:

16 oz	×	0.813	=	13 oz
Purchase unit		Yield		EP

One cup of cleaned green pepper weighs 5.2 ounces, so 1 pound (AP) of green peppers yields 2.5 cups:

13 oz	÷	5.2 oz	=	2.5 c per lb
Yield per pound		Weight per cup		Yield per pound

Therefore, the recipe requires 1.2 pounds of green pepper (AP), for a total ingredient cost of $3.66:

3 c	÷	2.5 c per lb	=	1.2 lb
Amount needed		Yield per pound		Amount to purchase

1.2 lb	×	$3.05	=	$3.66
Purchase amount		**Cost per pound**		**Ingredient cost**

- **Item L (bread crumbs):** 2 quarts are needed, the PU is pounds, and the cost per PU is $1.90. A table of edible food yields indicates that there are 4.6 cups of bread crumbs per pound. The recipe requires 8 cups (2 quarts), so 1.75 pounds are needed for an ingredient cost of $3.33:

8 c	÷	4.6 c per lb	=	1.75 lb (rounded)
Amount needed		**Volume per pound**		**Purchase amount**

1.75 lb	×	$1.90	=	$3.33
Purchase amount		**Cost per pound**		**Ingredient cost**

- **Item M (sharp cheddar cheese):** 2 quarts are needed, the PU is pounds, and the cost is $5.25 per pound. A table of edible food yields indicates that there are 4 cups of shredded (grated) cheddar cheese per pound. The recipe requires 8 cups (2 quarts), so 2 pounds of cheese are required for an ingredient cost of $10.50:

8 c	÷	4 c per lb	=	2 lb
Amount needed		**Yield per pound**		**Purchase amount**

2 lb	×	$5.25	=	$10.50
Purchase amount		**Cost per pound**		**Ingredient cost**

Each ingredient in the standardized recipe shown in *Exhibit 2.11* has now been costed.

While some time was needed to cost the recipe, the benefits, including knowledge of product costs for menu item pricing decisions, are worth the effort. Experienced managers can perform the calculations quickly. Also, computerized costing expedites the process and better ensures that the calculations are accurate.

Costing Standardized Recipes

After the ingredient costs in the standardized recipe are known, the total food cost to produce the recipe is easily calculated. That amount, shown at the bottom of column 6 in the Recipe Costing Worksheet (*Exhibit 2.13*), is $96.17.

In addition to the total ingredient cost of the standardized recipe, the food cost to produce one serving can be easily calculated:

$96.17	÷	48	=	$2.00
Total recipe cost		**Recipe yield as no. of servings**		**Per-serving cost**

MANAGER'S MATH

You are costing recipe ingredients. How much will the following ingredients cost?

1. A recipe requires 6 cups of trimmed and cleaned onions. Use the following information:
   - 1 pound of onions costs $1.45.
   - 1 cup of cleaned and trimmed onions weighs 4 ounces.
   - 1 pound of onions weighs 14.5 ounces after they are cleaned and trimmed.
2. A recipe requires 2 cups of cheddar cheese. Use the following information:
   - 1 pound of cheese costs $4.75.
   - There are 4 cups of shredded (grated) cheese per pound.

(*Answers: 1. Onions, $2.40; 2. Cheese, $2.38*)

OPEN FOR BUSINESS
RESTAURANT TECHNOLOGY

The advantages to costing standardized recipes depend on more than simply following the costing process. Three important requirements are necessary to get the full benefit:

- The standardized recipe must be used.
- Ingredients must be accurately weighed, measured, or counted.
- The standardized recipe must be costed using current purchase costs.

The ingredient costs in *Exhibit 2.13* will likely change over time, and then the costing process must be repeated. When a manual costing system is used, every recipe containing mushrooms, for example, must be revised to reflect a new cost. When a computerized system is used, all recipes containing mushrooms will be revised when the change is entered.

One serving of chicken tetrazzini is calculated to cost $2.00. However, remember that this will be the actual serving cost only if recipes are carefully followed, if ingredients are carefully weighed and measured, and if the recipe was costed with current market costs.

Calculating Menu (Plate) Cost

Some restaurants and foodservice operations use an à la carte pricing system in which entrées, soups, salads, and other menu items are priced individually. Customers "build" a meal when they place orders or select items from the serving counter. With this pricing plan, the food cost for each menu item, determined by recipe costing, can be the basis for menu pricing decisions.

More frequently, however, food items prepared with different recipes are available as a complete meal. An establishment may offer entrée items at different prices and **accompaniments** such as salad, potato, and other choices included as part of the entrée selling price. Guests may be offered a choice in some accompaniments, such as a selection from among three types of potato preparations; for example, French fries, baked potato, or garlic-mashed potatoes (*Exhibit 2.14*).

Exhibit 2.14

When this pricing plan is used, the cost to produce one serving of a specific food item must be combined with the costs of other food items to determine the total food cost for meals. **Menu costing**, also called plate cost, is the process of determining the food cost to produce all menu items that make up a meal offered at a set selling price when standardized recipes are used.

Exhibit 2.15 shows how the total of all per-serving costs for menu items in a specific meal is tallied. The total food cost for the Fresh White Fish Dinner will be $6.89 if the following factors are met:

- All meal components are prepared, portioned, and served according to standardized recipes.
- Each recipe has been costed with current ingredient costs.

Exhibit 2.15

MENU ITEM COSTING

Entrée: Fresh White Fish Dinner

Costing Date: 8/03/2012

Item	**Menu Item**	**Cost per Serving**
Entrée	Fresh white fish	$4.23
Potato	3 choices daily	0.37
Vegetable	4 choices daily	0.42
Salad	3 choices daily	1.12
Dressing	5 choices daily	0.37
Garnish	Lemon wheels	0.02
Bread loaf		0.27
Butter/margarine		0.06
Condiment(s)		0.03
	Total Entrée and Accompaniments Cost:	$6.89

Exhibit 2.15 indicates that guests selecting the Fresh White Fish Dinner are offered a choice of three potatoes, four vegetables, and three salads. Customers desiring a tossed green salad, which is one of the salad choices, have a choice of five dressings. There is also a lemon wheel garnish, and customers receive a small loaf of bread with butter or margarine, and condiments as desired.

How did the manager calculate the costs of potato, vegetable, salad, and dressing choices, since each likely has a different food cost and guests will order different accompaniments? The most conservative approach is to use the highest-cost choice in each category.

In this example, a serving of twice-baked potatoes has a higher cost than other potato choices. Its cost could be used for costing all potato choices. In *Exhibit 2.15* that cost is shown as $0.37. The same strategy could be used to determine the cost of the vegetable, salad, and salad dressing choices.

What happens when, for example, the purchase cost of a vegetable choice such as broccoli or sautéed mushrooms increases? There are several alternatives:

- Replace the vegetable choice with another that is within food cost limits.
- Retain the vegetable choice on the menu but price it on an à la carte basis and supplement it with another vegetable choice within cost limits.
- Serve a smaller serving size of the vegetable.
- Raise the food cost limit allowed for the vegetable choice.

COMPLETING RECIPE DEVELOPMENT REQUIREMENTS

Recall that the process for developing a standardized recipe entails eight steps. The previous information covers the process for the first six steps. In this section, information about the final two steps is provided.

Develop Food Safety Requirements

Step 7 in the standardized recipe development process involves identifying food safety requirements applicable to the recipe. When these are known, caution statements can be included in the recipe and important information for training and coaching will be available. **Coaching** is the process of encouraging employees to follow work practices they have been taught. A supervisor who observes an employee violating food safety practices while handling ingredients or prepared menu items can reinforce the proper procedures.

A comprehensive review of the recipe in line with the operation's **Hazard Analysis Critical Control Point (HACCP) system** is important. The HACCP system is used to control risks and hazards throughout the flow of food.

Exhibit 2.16 reviews HACCP principles and provides an overview of how each helps managers incorporate food safety concerns into operating procedures including those in recipes.

Exhibit 2.16

THE SEVEN HACCP PRINCIPLES*

1. Conduct a hazard analysis. 2. Determine critical control points (CCPs).**	These principles help identify and evaluate hazards.
3. Estimate critical limits. 4. Establish monitoring procedures. 5. Identify corrective actions.	These principles help establish ways to control the hazards identified.
6. Verify that the system works. 7. Establish procedures for record keeping and documentation.	These principles help maintain the HACCP plan and verify that it is effective.

*More information about HACCP can be found in the National Restaurant Association *ServSafe Coursebook.*

**Critical control points are points in an operation's foodhandling process in which identified hazards can be prevented, eliminated, or reduced to safe levels.

As *Exhibit 2.16* demonstrates, properties using a HACCP system have already identified and evaluated hazards, established ways to control them, and are maintaining the HACCP plan in a way that verifies it is effective.

Part of the ongoing maintenance and verification process is to determine the ingredients, if any, in a new recipe that should be under the HACCP plan.

Assume a recipe contains chicken and the managers recognize that bacteria are a likely hazard for chicken. They have previously developed proper handling processes for chicken throughout its flow through the operation and have determined that proper cooking is a step that will eliminate or reduce bacteria to safe levels. When the new recipe is standardized, it will contain the cooking time and temperature required to kill or reduce the bacteria to a safe level.

Production employees have been trained to properly handle chicken, and the procedures they learned apply to a new recipe containing this ingredient:

- The temperature of the chicken must be checked after cooking to ensure it has reached the proper internal temperature. If the chicken has not reached the proper internal temperatures, staff have been taught what to do.
- Ongoing checks are required throughout the shift to ensure that critical limits were met and appropriate corrective actions were taken with all applicable products, including chicken in the recipe being developed.
- A time–temperature log and receiving invoices are kept for a specified amount of time to further ensure that background information about chicken will be available if there is a problem. The time–temperature logs provide a history of times and temperatures at which the chicken was maintained during preparation and production. Receiving invoices confirm how long the chicken was on-site before it was served.

Some food safety information might be incorporated into standardized recipes. For example, assume the recipe is for beef stew, and the operation's policy is to reuse leftover portions. A recipe notation might be added that at the end of the serving period, the remaining product should be quickly refrigerated in small batches with occasional mixing to hasten the cooling process.

Implement and Consistently Use Recipe

Step 8 in the standardized recipe development process begins when the recipe is implemented. Standardized recipes should be followed every time a food item is prepared. However, in some operations they are not used except as a general guide. When this occurs, the creative efforts and time spent on developing the recipes are wasted. Even worse, the recipes cannot provide the consistent quality of products that customers demand and the consistent cost standards that the operation needs.

MAKE RECIPES AVAILABLE

Most operations have multiple hard copies of their standardized recipes. A master copy may be kept in the food-production manager's office with copies as needed in work areas. Some recipes are printed on thick paper stock to make them durable; others are on laminated paper so they can be wiped clean. Others are computer generated, with ingredient quantities and number of servings changed daily based on forecasts.

Some properties do not use hard copies of recipes and instead have them available on computers in the work area. This system is frequently used in bar operations so bartenders can make drinks that are not ordered frequently.

Recipes can be housed in recipe boxes, three-ring binders, folders of various types, or even in a drawer. Managers find the best ways to store and maintain them for their properties so they can be quickly retrieved when necessary.

PROVIDE TOOLS AND EQUIPMENT

Professional food-preparation staff understand that there is much more to using a standardized recipe than to "just do what it says." When a new recipe is introduced, it should be explained in detail. The employees who will use the recipe should go over it together, ensure that everyone understands what each procedure means, and agree on the weighing and measuring tools to be used.

While food-preparation staff members must carefully follow standardized recipes, their managers and supervisors must ensure that the required ingredients and tools are always available. If not, challenges will arise when desired items cannot be produced. Production problems can occur ("What can we prepare as an alternative item?"), and customers may be disappointed ("I came here just for that item!"). Also, quality and cost concerns arise if different or more expensive ingredients are used to produce the menu item.

Small wares, including pots and pans, specified by standardized recipes must be available and equipment must be in order. What if a recipe calls for baking a meat entrée at a specified temperature, but the oven thermostat has never been calibrated? **Calibrate** means to check or verify. The thermostat of an oven should be calibrated on a routine basis to confirm that the internal temperature is that for which the thermostat has been set.

Required measuring utensils must be provided. Why indicate that a number 8 scoop, which produces eight level servings per quart, be used for serving if this tool is broken or unavailable?

Some operations use a 1-gallon measure because this tool provides lines or ridges to indicate each quart amount. Perhaps more commonly, however, production managers provide tools for different volume measurements including gallon, quart, pint, cup, tablespoon, teaspoon, and even fractions of a teaspoon.

Most scales show measurements in both U.S. and metric weights so they can be used to measure items with either weight system. Many operations have at least three scales: a receiving scale that may weigh products up to 100 pounds or more, and two production scales, one that weighs quantities up to 25 pounds or more and a second that weighs ounces and fractions of an ounce up to, for example, 32 ounces (2 pounds).

Professional kitchen managers recognize the importance of calibrating scales for accurate weights just as they calibrate cooking equipment for accurate temperatures. This can easily be done by placing a container of known weight on a scale. For example, a 5-pound bag of sugar should obviously weigh 5 pounds. If the scale does not register this weight, calibration is required.

Many scales have a **tare allowance** feature that excludes the weight of a pot or pan placed on the scale to hold ingredients being measured. The pot or pan can be placed directly on the scale and the dial rotated to zero. If 3 pounds of flour are required, the food-preparation employee then needs to scoop flour into the pot or pan only until the scale reads 3 pounds.

Without a tare allowance feature, the pot must be placed on the scale and its weight must be noted. Then 3 pounds of flour must be added to that weight. If the pot weighs 1 pound 2 ounces and 3 pounds flour are required, flour must be added until the scale reads 4 pounds 2 ounces. In a busy, noisy kitchen, simple features such as a tare allowance scale make it much easier to follow standardized recipes accurately.

Weighing is always more accurate than measuring by volume. Even though most people would probably measure water by volume, many professional bakers weigh the water for their recipes. Instead of a baking recipe specifying 2 cups of water, it will require 16 ounces of water, which is the weight of 2 cups.

> **Manager's Memo**
>
> An old saying states, "A pint is a pound the world around." It would be easy to convert measures to weights if this statement were always true. Unfortunately, it is not, so kitchen tools such as scales and measuring utensils are needed.
>
> It is true that a pint measure (2 cups) holds 1 pound (16 ounces) of water. However, the same pint measure may hold only 8 ounces of flour. The actual weight depends on, among other things, whether the flour has been sifted or packed into the pint measure.
>
> Use the Internet to find sites that can help with weight and measure equivalents. Type the phrase "weight and measure equivalents" into your favorite search engine.

TECHNOLOGY AND STANDARDIZED RECIPES

Increasingly, food-production managers use computers and advanced software programs to calculate recipe costs, estimate ingredient purchase quantities and costs, and perform other purchasing functions. If standardized recipes are not used, these tasks cannot be performed electronically because they require current data. This is very time-consuming to generate without standardized recipes. As a result, the ability to use advanced technology tools is severely reduced for operations that do not use standardized recipes.

Computer-Generated Recipes

Automated systems are increasingly being used to adjust recipes for the specific number of servings to be prepared each day (or other time period). Consider Murray's Seaside Restaurant, which is known for its seafood casserole. At the end of each day, the food-production manager determines the quantity of seafood casserole remaining because it can be re-served the next day. She also estimates the number of servings to be sold the next day and makes a simple calculation:

No. of servings needed next day	−	No. of servings available end of shift today	=	No. of servings to be produced next day

The manager's recipe system is computerized. As a result, when the number of servings required for production the next day is entered into the system, a recipe is generated that indicates the quantity of each ingredient needed for that number of servings.

Each morning when production employees check in for work, they receive the recipes adjusted for the number of servings of each menu item they should produce. This system allows the operation to minimize food costs and maximize food quality while "fine-tuning" how sales are estimated to minimize leftovers.

Computerized Costing

The procedures and types of calculations needed for recipe costing are the same whether the process is done manually or with a computerized system. Technology increasingly impacts many aspects of restaurant and foodservice operations, and this includes planning for food costs before they are incurred. Common **recipe management software** (computer programs that involve or impact standardized recipes) frequently maintains three files that help control food costs.

- **Ingredient file:** This computerized record contains information about each ingredient purchased, including PU size and cost, issue unit size and cost, and recipe unit size and cost. Each of these units can differ. Prepared applesauce can be purchased in a case of six #10 cans (purchase unit), issued to the kitchen one can at a time (issue unit), and used in recipes by the cup (recipe unit).

  Much time is normally required to develop and maintain an ingredient file for each ingredient purchased. With some computerized systems, the ingredient file is available through inventory software that maintains information about the quantity and cost of items in current inventory and which have been issued. The ingredient file must maintain current and accurate information because its data drive the standardized recipe and menu item files.

- **Standardized recipe file:** This computerized record contains the recipes for menu items produced. Data include each recipe's ingredients, preparation methods, yield (number of servings and serving size), and ingredient costs along with each item's selling price and food cost percentage (Food cost ÷ Selling price).

Manager's Memo

Food-production staff should always follow standardized recipes when they prepare food products. Bartenders should also use standardized recipes for the beverages they prepare. Sometimes bartenders pour spirits into drinks without accurately measuring them. They reason that they are experienced and too busy to use a measuring tool.

Hopefully, food-production staff members do not have this attitude. The comment "I know how much 3 pounds of rice weighs" would never come from a professional food-preparation employee who knows about the need for careful weighing and measuring.

Consistency, an advantage of standardized recipes, is possible only when ingredients are accurately weighed or measured and when preparation procedures are carefully followed.

Recipe conversion software can convert recipe ingredients from weight to volume measurements or vice versa and then calculate the ingredient costs for the recipe.

- **Menu item file:** This computerized record contains information about menu items tracked with the operation's **point-of-sale (POS) system**. The POS records an operation's sales, product usage, and other important information on a daily, by shift, hourly, or other basis (*Exhibit 2.17*). The menu item record in the POS would include the menu item's selling price, ingredient quantities, and unit sales totals.

Exhibit 2.17

SUMMARY

1. **Explain the importance of standardized recipes.**

   Every restaurant or foodservice operation needs standardized recipes. They provide consistency for customers, so menu items will always look and taste the same and provide the same value. They also provide consistency for the operation to ensure that the selling price is based on known ingredient costs.

2. **List the procedures for developing standardized recipes.**

   When recipes are being standardized for menu items currently produced, managers observe, consider preparation details, and write, review, and revise a recipe draft. Then the recipe can be used and the item evaluated to determine if further revisions are necessary. Finally, the recipe should be implemented and consistently used.

   With recipes for new menu items, a slightly different procedure is required. First, recipe alternatives should be collected, a rating scale developed, and a taste-test panel used to evaluate the recipes. Then the recipe judged "best" should be considered against operating concerns.

3. **Describe how to standardize recipes for a specific operation.**

   Recipe adjustments may be needed to the number of servings and serving sizes. In each case, a recipe conversion factor (RCF) is calculated. The amount of each ingredient is multiplied by the RCF to determine the quantity required for the converted recipe.

   Cooks in the United States must know volume measures from gallon to teaspoon, cup to fraction of a teaspoon, and weight in pounds to ounces. Some recipes may need to be converted from metric measurements.

   A format for the standardized recipe should be agreed upon and designed to provide all information necessary about ingredients, quantities, and preparation.

OPEN FOR BUSINESS — RESTAURANT TECHNOLOGY

Some time is required to update ingredient files when purchase prices change. However, managers will spend significantly less time maintaining current ingredient costs with an electronic system compared to performing these calculations manually.

For example, Murray's Seaside Restaurant has five recipes using canned tomato sauce. When the cost of this ingredient changes, the production manager must only change the price once in the ingredient file. Then the new purchase cost will carry over and adjust each affected standardized recipe accordingly. Trends seem to indicate that the use of automated software to control food costs will increase, and recipe management software is among the elements that will likely be more commonly used.

4. **Calculate recipe ingredient costs, standardized recipe costs, and menu (plate) costs.**

   Recipe costing considers the cost to produce all or a single serving of a recipe by calculating ingredient costs and the number of servings the recipe yields. Tables of edible food yields are needed to determine the usable amount of a food ingredient when trimming is involved or weight or volume conversion is needed.

   After ingredient costs are determined, they can be added together to calculate the recipe cost. This cost, in turn, can be divided by the recipe's yield to determine the cost per serving.

   When establishments include accompaniments with the entrée for a single selling price, serving costs for each item must be added to determine the total food cost for the meal.

5. **Explain how to ensure food safety requirements are addressed in standardized recipes, and explain procedures helpful when implementing and consistently using standardized recipes.**

   Many operations use a Hazard Analysis Critical Control Point (HACCP) system to develop procedures for managing identified food hazards. These procedures should be incorporated into standardized recipes.

   Numerous details must be addressed as menus are implemented, from calibrating ovens and scales to accurately weighing and measuring ingredients. Food-preparation staff must always follow the standardized recipe carefully.

6. **Describe how technology impacts the use of standardized recipes.**

   Electronic systems can generate recipes daily (or more frequently) that have been adjusted for the specific number of servings to be prepared. Recipe management software can automatically update ingredient costs when market prices change.

APPLICATION EXERCISES

Exercise 1: Recipe Evaluation

Review *Exhibit 2.4* (Standardized Recipe Evaluation Form). Think about several of your favorite menu items. Review the evaluation factors in *Exhibit 2.4* and list any additional evaluation factors that would be important. Also, note any evaluation factors that would not be applicable.

Name of Favorite Item	
List of New Evaluation Factors	
Evaluation Factors That Do Not Apply	

Exercise 2: Recipe Adjustments

Part A

A recipe yields 75 servings (4 oz), and you want to adjust it to yield 50 servings (4 oz). One of the ingredients is flour: six oz are needed for the current recipe. How much flour will be needed for the converted recipe?

Part B

A recipe yields 75 servings (¾ c), and you want to adjust it to yield 75 servings (½ c). One of the ingredients is fluid whole milk; 1/4 c is needed for the current recipe. How much milk will be needed for the converted recipe?

Part C

A recipe yields 60 servings (4 oz), and you want to adjust it to yield 75 servings (5 oz). One of the ingredients is sugar; ½ c is needed for the current recipe. How much will be needed for the converted recipe?

Exercise 3: Recipe Costing

Your operation sells a lot of cheeseburgers. Calculate the food cost given the following information:

- Ground beef: 6 oz portion (AP) purchased at $4.10 per lb.
- Cheese: 1.5 oz purchased at $5.25 per lb.
- Lettuce: 1 leaf (assume 20 leaves per head) purchased in a case of 24 heads at $27.50 per case.
- Onion: 1 slice (onion yields 12 slices); 2.5 onions in a pound purchased at $1.74 per lb.
- Bun: 1 bun in package of 18 purchased at $3.20 per package.
- Tomato ketchup: 3 oz average purchased in gallon container (128 oz) packed 4 gal per case for $28.05.

REVIEW YOUR LEARNING

Select the best answer for each question.

1. **Which is a benefit of standardized recipes?**
   A. Consistency is provided for customers and the operation.
   B. The operation can purchase less expensive ingredients.
   C. Kitchen employees can use time-saving procedures.
   D. There will be a minimum of leftover food.

2. **What is the first step in developing recipes for menu items currently produced?**
   A. Assign one menu item to each cook.
   B. Develop recipes for the most difficult items first.
   C. Tell food-production staff why recipes are needed.
   D. Ask regular customers for advice on changing recipes.

3. **Edible portion (EP) refers to the amount of food that**
   A. is received from suppliers.
   B. can be served after cooking.
   C. remains after initial processing.
   D. can be re-served the next day.

4. **A recipe yields 25 servings (3 oz), and 70 servings (3 oz) are needed. What is the recipe conversion factor (RCF)?**
   A. 2.1
   B. 2.8
   C. 3.1
   D. 3.6

5. **How many quarts are in 1½ gallons?**
   A. 2
   B. 4
   C. 6
   D. 8

6. **How many ounces are in ½ pound?**
   A. 4
   B. 6
   C. 8
   D. 10

7. **One pound of cheese costs $6.25. A standardized recipe requires 4 oz of cheese. What is the approximate cost of the cheese required for the recipe?**
   A. $3.15
   B. $2.40
   C. $1.56
   D. $0.85

8. **What does the term *accompaniments* refer to?**
   A. Food items on an all-you-can-eat salad bar
   B. Condiments or garnishes on a dinner plate
   C. Choices included as part of an entrée price
   D. Items on a dining table such as salt and pepper

9. **A tare allowance feature on a scale**
   A. excludes the weight of a pot placed on the scale.
   B. converts U.S. to metric system measurements.
   C. allows production employees to measure liquids.
   D. enables cooks to weigh items in fractions of an ounce.

10. **Information on what document should be available if there is a concern about the use of food safety procedures as a menu item is prepared and served?**
   A. Purchase order
   B. Issue requisition
   C. Closing checklist
   D. Time–temperature log

FIELD PROJECT

1. When opening a new restaurant, it is important that a standard format be used for the recipes. Visit a favorite establishment and ask the manager if you can have a copy of a standardized recipe that is used at the establishment. Alternatively, you may ask a foodservice manager at your school for a recipe or use your favorite search engine and type "standardized recipe format" to view numerous examples.

   Use the sample you obtain and compare it to the standardized recipe format shown in *Exhibit 2.11*. Then develop a recipe format that incorporates the best features of both recipes.

2. Use the Internet to search for a possible standardized recipe for an entrée that might be useful for the Italian-themed establishment you are planning. Transfer the information to the standardized recipe format you will use in your new operation.

3

Product Purchasing

INSIDE THIS CHAPTER

- Importance of Purchasing
- Determining Quality Requirements
- Determining Purchase Quantities
- Selecting Vendors
- Product Ordering Procedures
- Purchasing Follow-Up

CHAPTER LEARNING OBJECTIVES

After completing this chapter, you should be able to:

- State the objectives of an effective purchasing process.
- Define quality, and review purchasing procedures that help ensure that operations obtain products of the proper quality.
- Describe basic procedures for determining the quantity of products to purchase.
- Identify concerns important when selecting product vendors.
- Explain basic product ordering procedures.
- Describe basic payment methods and the management of credit memos and petty cash funds.

KEY TERMS

"A" item, p. 80
broad line vendor, p. 86
capital, p. 77
cash flow, p. 77
centralized purchasing, p. 71
close-out, p. 77
competitive bidding, p. 71
decentralized purchasing, p. 71
expediting, p. 72
high check average, p. 73
minimum–maximum inventory system, p. 80
net price, p. 89
ordering, p. 87
order period, p. 80
par inventory system, p. 83
perpetual inventory system, p. 79
petty cash, p. 93
petty cash voucher, p. 96
physical inventory system, p. 78
purchase order (PO), p. 72
purchase requisition, p. 71
purchase specification, p. 73
purchase unit (PU), p. 80
purchasing, p. 70
request for proposal (RFP), p. 71
safety level, p. 80
specialty line vendor, p. 86
stock out, p. 77

CASE STUDY

"This is the second time it happened this week!" exclaimed Rick, the food-production manager at the Blue Haze Grill. "Our seafood vendor delivered some seafood that wasn't up to our standards."

"What did you do about that?" asked Andy, the beverage manager, during a department head meeting.

"Well," said Rick, "We accepted the product because we had no other choice, but we had to make a special dish to offer in place of the regular menu item. We are also looking for another vendor who knows how to get us the quality of product we need."

1. What should the management team at the establishment do to resolve this problem?
2. What should Rick say to the vendor, and what can the vendor do to address the problem in the future?

IMPORTANCE OF PURCHASING

Purchasing is the series of activities that begins when needs are determined by the menu, and ends after products are served. It includes an emphasis on vendor interactions. This definition is broader than the common idea that purchasing simply means buying: determining purchase quantities, placing orders, and paying vendors. Experienced restaurant and foodservice professionals know that much planning is required even before orders are placed.

The people responsible for purchasing must identify and obtain the products and services that allow their organization to meet the expectations of their customers at a price that provides value. The process is never-ending because customers' preferences change, new product alternatives are continually introduced, and cost concerns always require attention.

Objectives of Effective Purchasing

Purchasing objectives are simple to state, and they are reviewed in *Exhibit 3.1*. However, attaining these objectives is much less simple. As *Exhibit 3.1* demonstrates, when purchasing objectives are met, both the customers and the operation benefit.

Exhibit 3.1

PURCHASING OBJECTIVES

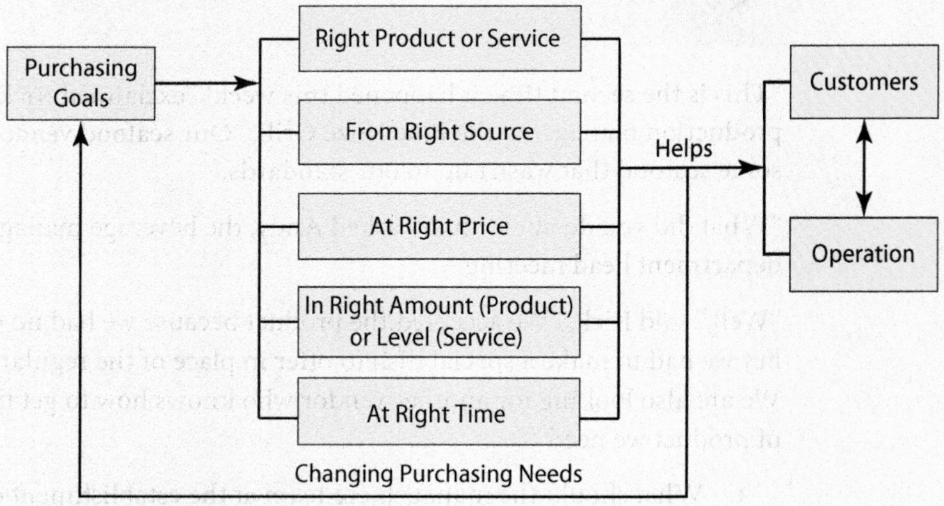

Purchasing objectives drive what managers must do. For example, if an objective is to obtain products at the "right" price, then procedures must be developed that can be used to obtain and evaluate prices from approved vendors for products meeting quality requirements.

Managers must help determine the needs for specific products and identify the best sources for them. The objectives also influence the manager's day-to-day responsibilities, which include the following:

- Ensuring that products are available when needed
- Developing purchasing policies and procedures
- Using purchasing practices to yield the best value for the operation and its customers
- Ensuring on-time vendor payments and maintaining good vendor relationships

Purchasing Responsibilities

Large restaurant and foodservice organizations typically have a purchasing director, and may require an entire purchasing department. They use a **centralized purchasing** system in which purchasing requests are routed to those with specialized responsibilities, who then purchase the products.

In small organizations, persons with many responsibilities perform purchasing tasks, and they often use a **decentralized purchasing** system to do so: department heads may do the purchasing, often with input and approval by the manager, especially for high-cost or large quantity purchases.

Those with purchasing responsibilities must do many tasks, regardless of the size of their operation. Some of these must occur before the initial contact with vendors:

- Determine products to purchase.
- Develop purchase specifications that indicate the quality requirements for products that will be purchased.
- Review **purchase requisitions**: requests to purchase products required by departments.
- Determine correct purchase quantities.
- Determine the purchase method such as direct purchase or **competitive bidding**: the comparison of vendors' prices for products of acceptable quality to determine the least expensive alternative.
- Prepare **requests for proposals (RFPs)** to request prices from vendors for products of a specified quality.

Other purchasing activities require direct interactions with vendors:

- Interview possible vendors and collect and analyze information to determine which among possibly many vendors will be sent RFPs.
- Communicate with vendors to obtain product information.
- Send RFPs to approved vendors.

THINK ABOUT IT . . .

Managers of small operations are responsible for purchasing activities and many other challenges that require specialized knowledge.

How can a manager know how to do almost everything done by different specialists in a large operation?

- Review vendors' price proposals, negotiate contracts, and select vendors for specific orders.
- Send **purchase orders (POs)** to vendors to inform them that their proposals were accepted and that the orders should be delivered.
- Expedite product deliveries. **Expediting** refers to interacting with vendors about POs that have been placed but for which products have not been delivered.
- Use proper product receiving practices to accept product deliveries including signing delivery invoices, which transfer ownership of products from the vendors to the operation.
- Take corrective actions when delivered products do not meet standards.

Still other purchasing activities involve after-delivery payment and ongoing responsibilities:

- Ensure purchasing documents are routed for payment.
- Keep current with forecasted product costs, availability, and new products.
- Manage purchasing policies and procedures.
- Review, revise, and maintain product purchase specifications.
- Resolve vendor problems.
- Maintain a vendor handbook.
- Maintain manual or computerized purchasing records.

DETERMINING QUALITY REQUIREMENTS

The old saying "You get what you pay for" is generally true in purchasing. Value is the relationship between price and quality, and it is very important. Managers seek value as they make purchasing decisions in the same way customers do when they purchase menu items. Wise managers know that savings from effective purchasing can be passed on to customers. This encourages greater value perception and helps the operation attain financial goals.

What Is Quality?

Some buyers make a decision about quality when they see products delivered by a vendor (*Exhibit 3.2*). Then the only way to resolve a quality problem is often to reject inferior items, which can disrupt production, or to negotiate a lower price—and a lower price for a product of inadequate quality is never a "deal." However, effective managers plan and communicate with vendors to best ensure that quality standards will be met when products are received.

Exhibit 3.2

Product quality relates to a product's intended use. The buyer must define what a product will be used for and then select a suitable quality of product. If an operation requires tomato products for spaghetti, what market form should be used: fresh whole tomatoes, tomato pieces of uniform or irregular sizes, tomato sauce, or tomato puree? If the property is a **high check average** operation where high selling prices offset the labor costs needed to process fresh food items, fresh tomato products may be used in the spaghetti sauce. However, if the facility offers much lower selling prices, processed tomato products may be used. At some operations, fresh tomatoes may be used for almost every product requiring tomatoes. At others, they might be used only for fresh salads and plate garnishes.

Purchase Specifications Define Quality

Purchase specifications describe the quality requirements of the products that are purchased. They help ensure that user department staff, managers, and vendors have the same understanding about quality needs.

Effective purchase specifications must meet several requirements:

- They must be simple and short while providing accurate product descriptions.
- They must be capable of being met by several vendors to encourage competitive bidding.
- They must be capable of being verified to ensure that proper quality is received.
- They must be written to permit reasonable compliance with quality requirements.

Manager's Memo

There are three basic types of managers:

- *Price-conscious managers* who are primarily concerned about price and whose purchasing goal is to buy the least expensive products.
- *Quality-conscious managers* who are not overly concerned about costs. Instead, they want to purchase products of the highest possible quality and believe the greater costs can be passed on to customers.
- *Value-conscious managers* who recognize the relationship between a product's price and its quality. They aim for neither the lowest or highest *price* nor the lowest or highest *quality*. They know the right product is the one that represents the greatest value based on its suitability for intended use.

THINK ABOUT IT . . .

Purchase specifications take time to develop. One cost-effective approach is to develop specifications only for expensive products and products purchased in large quantities. What do you think of this "compromise" approach?

Purchase specifications inform vendors about the operation's required quality standards. They allow vendors to quote prices for applicable products when quantity and other order requirements are known. Specifications can be described in several ways:

- **Brand or trade name:** Managers who specify a specific brand of ketchup are indicating a quality preference.
- **Certification with industry specification:** Consider the North American Meat Processors Association (NAMP) and its Meat Buyer's Guide (MBG) numbers. When a manager specifies, for example, NAMP #1185, a bottom sirloin butt steak meeting specific quality standards is identified. Federal grading standards are additional examples of widely circulated and recognized specifications available for many products.
- **Careful description of required products:** There are many products, such as fresh seafood, meat, and dairy products, that cannot be identified by a brand name or a trade-recognized product number. Written purchase specification statements are needed for these items.
- **Use of samples:** Samples of products currently used by an operation may serve as specifications. For example, a manager may want to buy ready-made specialty bread products, and samples prepared by the establishment may be provided to possible vendors for evaluation.

DEVELOPMENT OF SPECIFICATIONS

Managers must facilitate the development of purchase specifications by learning about the products that are needed to prepare menu items. They do so as they read industry trade publications, meet with vendors, and attend trade shows. They can discuss their needs with existing vendors and others in their professional network. Staff members with extensive industry experience may be able to suggest potential supply sources, and they may have their own network of peers who can make suggestions.

After potential vendors are identified, managers can obtain information and product samples for evaluation. They can facilitate in-house analysis of products currently used for which specifications are being developed or revised and of new products for which specifications do not exist. Meetings including taste panels with department staff, other managers, and even customers can be conducted.

The employees who will use the product will know about its intended use. Additionally, they will likely have ideas about positive and negative features of existing products, and they can identify important quality characteristics.

The next step is to develop draft versions of the specifications. General information from vendors and trade organizations along with specific characteristics suggested by in-house staff can be useful. Managers can then share drafts of proposed purchase specifications with vendors to obtain feedback and to ensure that the resulting specifications do not limit the number of vendors who can provide products. For example, use of a brand name in a specification may limit the number of vendors who can supply a product. Or, vendors may have suggestions about the types of products useful for self-service salad bars and food bars. Managers will benefit from the technical information and support provided by vendors during the specification development process.

THINK ABOUT IT . . .

In the complex world of purchasing, the advice of specialists can be helpful. Vendors are experts, and their assistance in developing specifications provides value.

What types of information can vendors provide to managers?

SPECIFICATION INFORMATION

Many operations use a template of basic information to develop their purchase specifications. An example is shown in *Exhibit 3.3.*

Exhibit 3.3

PURCHASE SPECIFICATION FORMAT

Product name: ______________________ Specification no.: __________

Menu item name (if applicable): ______________________

Product use: ______________________

____

____

General product description: ______________________

____

____

Specific information (as applicable)

- Count/portion size: ______________________
- Processing requirements: ______________________
- Drained weight (canned items): ______________________
- Trade number or grade: ______________________
- Weight: ______________________
- Variety, style, type: ______________________
- Packaging requirements: ______________________
- Geographic origin: ______________________
- Yield percent: ______________________
- Other special information: ______________________

____

____

Quality inspection procedures: ______________________

____

Other requirements and general information: ______________________

____

Specification implementation date: ______________

Manager's Memo

It does no good to develop purchase specifications unless they are used. For example, vendors who will be asked for price quotations should receive a copy of applicable purchase specifications. Then the prices they quote should correspond to the quality requirements identified.

Current and accurate specifications identify the quality standards required. Therefore, copies should also be given to receiving staff to help them confirm that incoming products meet quality requirements. The specifications can be an excellent training resource to help new receiving employees learn about the operation's product quality requirements.

Exhibit 3.4

Exhibit 3.3 shows many details normally addressed in a purchase specification. However, it also includes two less common features: product use and quality inspection procedures (*Exhibit 3.4*).

Recall the definition of quality that considers a product's intended use. Product use information in a purchase specification is important to help vendors understand the purpose for which a product is needed. They might suggest other products not considered or not available when the specification was developed.

Including quality inspection procedures on the specification emphasizes that the operation is serious about ensuring that requirements are met. Also, most vendors want managers to receive products of the proper quality because they know that unethical competitors may substitute products of lower quality to quote low prices.

DETERMINING PURCHASE QUANTITIES

Managers must be concerned about buying too much or too little as they make purchase quantity decisions. Many operations have menus that change infrequently, so purchase needs do not change rapidly either. However, the quantity of products required for each purchase cycle can change because of different estimates of the number of customers to be served.

Managers typically use different procedures to determine purchase quantities for different products. Highly perishable items such as dairy products and fresh produce are typically purchased in quantities that will be used over a several-day period. Frozen and canned food items may be purchased for several months' usage, or longer. Other products, such as cleaning supplies and paper goods, could be purchased in quantities sufficient for even longer time periods.

Improper Product Quantities

Knowing the right amount of a product to purchase for a given time period is critical. Problems can occur when products are purchased in greater or lesser quantities than necessary.

EXCESSIVE QUANTITIES

Problems can arise when an excessive quantity is purchased:

- The purchase ties up capital, the amount of the owner's money invested in the business that could be used for other purposes.
- The purchase impacts cash flow, the amount of money needed to pay bills when due.
- More space must be available to store products.
- There is an increased risk of theft or even damages; for example, if products continue to be moved around to make way for other incoming items.
- Quality deterioration may occur with perishable products.
- Handling costs increase. For example, additional time is required to conduct inventory counts.

INADEQUATE QUANTITIES

Inadequate purchase quantities can create a stock out, or a situation in which a product is no longer available in inventory. This can create other problems:

- Inability to meet production requirements
- Need to revise production plans
- Possibility of disappointed customers
- Loss of sales and revenue

Factors Affecting Purchase Quantities

The primary factors related to purchase quantity decisions involve the sales forecast (which drives production needs), volume purchase discounts, storage space, and cash flow. Procedures for production forecasting are presented in chapter 5.

Several other concerns may be important when purchase quantities are determined:

- **Minimum orders:** Many vendors specify a minimum dollar value of products for delivery.
- **Anticipated increases or decreases in product prices:** When prices are increasing, products may be purchased in larger-than-normal quantities to take advantage of lower prices on current purchases. When prices are decreasing, operations may purchase in smaller quantities to take advantage of lower prices on future purchases.
- **Close-outs:** Larger quantities may be purchased when, for example, vendors or manufacturers offer close-outs: short-term promotional discounts to introduce new products or to quickly sell products with outdated packaging.

Inventory Levels Impacting Purchase Quantities

The quantity of products on hand impacts decisions about when and how much to purchase. All operations should conduct a physical inventory at least monthly, and many properties benefit from maintaining a perpetual inventory count, at least for expensive and theft-prone products.

Exhibit 3.5

PHYSICAL INVENTORY SYSTEM

A **physical inventory system** involves manually counting the number of each product on hand (*Exhibit 3.5*). This must be done to obtain cost of goods sold (food cost) information for the monthly income statement. Some managers require someone other than the person responsible for the inventory to conduct the physical count to help ensure that no control problems such as theft exist. *Exhibit 3.6* shows a physical inventory form that identifies the information that must be collected.

Note that two cases of green beans were in the storeroom when the inventory count was made. The delivery invoice price was marked on the case when received, and each has a purchase price of $31.50, so the total cost of the green beans is $63.00:

2	×	$31.50	=	$63.00
Total cases		**Cost per case**		**Cost price**

When the physical inventory process is complete, managers can calculate the monthly food costs and will have information helpful in determining whether additional product purchases should be made.

Exhibit 3.6

PHYSICAL INVENTORY FORM

Item	Purchase Unit	No. of Units in Kitchen Storage	Purchase Price	Total Cost
Green beans	Case	2	$31.50	$63.00
			Total	$503.00

PERPETUAL INVENTORY SYSTEM

A **perpetual inventory system** is a continuous count of the number of items in inventory. The key advantage of continually updating the record for every item is that the managers always know the quantity of products that should be available.

Managers use a perpetual inventory system just like a checkbook: as dollars (products) are deposited in the bank (brought into the storeroom), the balance in the account (inventory record) increases. As dollars are withdrawn (products are issued), the balance decreases. Managers know the amount of dollars (quantity of products) that should be in inventory at all times. *Exhibit 3.7* shows a perpetual inventory form.

Exhibit 3.7

PERPETUAL INVENTORY FORM

Item: Strip Steaks (6 oz)

Date	No. of Purchase Units		Balance
	In	Out	
			37
9/09	-----	25	12
9/10	35	20	27

Strip steaks are judged to be expensive and theft-prone, so they are included in the perpetual inventory system. There were 37 individual 6-ounce servings of this item available at the beginning of the period. On the first date (9/09), 25 steaks were issued, so only 12 steaks remained in inventory:

37	−	25	=	12
Steak beginning inventory		**Steaks issued on 9/09**		**Steaks remaining**

On the next day (9/10), 35 steaks were purchased and 20 steaks were issued. There was then a balance of 15 steaks, so the inventory balance increased to 27 servings:

12	+	35	−	20	=	27
Steak beginning inventory		**Steaks purchased on 9/10**		**Steaks issued on 9/10**		**Steaks remaining**

OPEN FOR BUSINESS — RESTAURANT TECHNOLOGY

Computerized systems are available to help with inventory counts and to determine inventory costs. For example, optical scanners can be used to read bar codes on containers in storage. This technology provides a fast and accurate method of determining inventory costs. When products enter storage, the quantities and costs of the inventory are automatically increased. When products are issued into production, the quantities and costs are automatically decreased.

Technology can help to quickly and continually determine the quantity of products available so that the quantities to purchase can be calculated. In the future, these or even more useful systems may replace the manual counting systems used in most operations today.

> **Manager's Memo**
>
> For reasons including convenience, cost, and practicality, most managers do not purchase individual products when the quantities on hand reach an ideal order point. Instead, items are typically divided into several categories, and all or most products of that type are ordered at the same time. Categories may include fresh produce, fresh meat, and dairy products. Orders are placed for all items in one of these categories at the same time (and often from the same vendor).

The manager should spot-check the number of strip steaks in the storage area to ensure that the number of servings available equals the balance on the perpetual inventory form.

MINIMUM–MAXIMUM INVENTORY SYSTEM

Products that will remain in storage for more than a few days can be managed with a **minimum–maximum inventory system**. This system indicates the minimum quantity below which inventory levels should not fall and the maximum quantity above which levels should not rise.

The minimum–maximum system is best used to control the relatively few and most expensive **"A" items**: those that cost the most.

The minimum–maximum inventory system is also best used in certain situations:

- Product prices are relatively constant.
- Products are used in relatively consistent quantities.
- The same type of products will be used in the future.
- Products are not perishable.
- Reasonable maximum quantities do not present storage space problems.
- Inventory and storage procedures help ensure that stock is rotated and theft is minimized.

Several terms are important for understanding the minimum–maximum inventory system:

- **Purchase unit (PU):** The standard size of the package or container in which the product is typically purchased. For example, many canned fruit and vegetable products are purchased by the case. There are six #10 cans per case.
- **Product usage rate:** The number of purchase units used during a typical order period.
- **Order period:** The time in days or weeks for which an order is normally placed. For example, canned goods may be purchased once monthly.
- **Lead time:** The number of purchase units typically used during the time between order placement and delivery. If three 50-pound cases of frozen shrimp are normally used between product order and receipt, the lead time is three cases.
- **Safety level:** The minimum number of purchase units that must always remain in inventory in case of late deliveries or unexpected increases in product usage rates.
- **Order point:** The number of purchase units to be available in inventory when an order is placed.

For example, if an operation uses a large quantity of frozen shrimp of a specified size, and the product is included in the minimum–maximum inventory system, managers know the following information:

- Purchase unit: 1 case = 10 boxes (5 lb of shrimp per box) = 50 lb of shrimp total
- Product usage rate: 42 cases per order period
- Order period: 2 weeks (14 days)

  Note: *Daily usage rate = 3 cases (42 cases per order period ÷ 14 days)*
- Lead time: 4 days × 3 cases per day = 12 cases

  Note: *Number of cases used during lead time = 12 cases (3 cases per day × 4 days)*
- Safety level: 12 cases

Managers can now answer several questions about the purchase quantities for frozen shrimp.

Question 1: What is the maximum number of cases of shrimp that should ever be available in inventory?

To answer this question, the manager adds the usage rate to the safety level. The usage rate is 42 cases and the safety level is 12 cases:

42 cases	**+**	**12 cases**	**=**	**54 cases**
Usage rate		**Safety level**		**Maximum**

Question 2: What is the order point for the shrimp?

The order point is the sum of the number of lead time cases and the safety-level cases:

12 cases	**+**	**12 cases**	**=**	**24 cases**
Lead time		**Safety level**		**Order point**

The order point can be verified because the number of cases available when the order is placed minus the number of cases that will be used before the order arrives is equal to the safety level, which is the minimum number of cases that can be available:

24 cases	**–**	**12 cases**	**=**	**12 cases**
No. of cases available when shrimp is ordered		**No. of cases used until shrimp is delivered (lead time)**		**No. of cases available when shrimp is delivered (safety level)**

Question 3: How many cases of shrimp should be ordered at the order point?

The usage rate (42 cases) should be ordered if the order is placed at the order point. This can be verified:

24 cases	−	12 cases	=	12
Order point		**Lead time**		**Cases at delivery**

12	+	42	=	54 cases
Cases at delivery		**Cases ordered (usage rate)**		**Maximum inventory cases**

Question 4: How many cases of shrimp should be ordered if an order is placed when there are 30 cases in inventory? In this example, the order point for shrimp has not been reached, but the order is placed with other frozen seafood products.

Step A: Calculate the number of cases of shrimp that exceed the order point:

30 cases	−	24 cases	=	6 cases
Cases in storage		**Order point**		**Excess**

Step B: Calculate the number of cases to order:

42 cases	−	6 cases	=	36 cases
Usage rate		**Excess**		**Order**

The number of cases to order when there are 6 cases in excess of the order point can be verified:

30	−	12	=	18
Cases in storage		**Lead time cases**		**Cases at delivery**

18	+	36	=	54
Cases at delivery		**Cases ordered**		**Maximum inventory cases**

The previous questions demonstrate that product usage rate, order period, lead time, and safety level can be used to determine key information:

- The minimum number of cases allowable in inventory
- The maximum number of cases allowable in inventory
- The number of cases that should represent the order point
- The number of cases that should be purchased if an order will be placed before that product's order point is reached

Two factors should be considered when establishing a product's safety level:

1. **The lead time required for reorders:** As the frequency of deliveries decreases, the number of safety-level units should increase.
2. **The product's usage rate:** As the volume of product usage increases, safety levels may need to be increased accordingly.

An ideal safety level minimizes stock outs without maintaining excessive quantities of products in storage.

Several factors influence decisions about product lead times. Managers develop lead times by considering the time between order and delivery. Some conditions increase the length of lead times:

- Vendors are not dependable. This should also be an important factor in deciding whether to do business with them.
- The operation is in a remote location, and delivery delays are common.
- Market situations cause unpredictable conditions that affect product availability and the potential for backorders.

There are potential advantages to using the minimum–maximum inventory system. First, excessive stock buildup is less likely when reasonable maximum inventory levels are established. Also, the minimum level provides a cushion against stock outs.

There are also potential disadvantages to using the minimum–maximum system:

- It may not be the optimal way to calculate required quantities. There are, for example, computerized systems that provide more detailed and accurate forecasts of purchase needs and timing.
- The assumptions used to establish the system's safety and lead-time calculations may not always be accurate.
- Time is required to calculate accurate safety and lead-time estimates.

PAR INVENTORY SYSTEM

A **par inventory system** also considers the quantity of products needed to bring the inventory level to an allowable maximum or "par." It uses a practical and simple method to calculate purchase quantities for perishable products including dairy and bakery products, and for other products such as packaged coffee and teas, alcoholic beverages, and cleaning chemicals with predictable usage rates.

OPEN FOR BUSINESS — MANAGER'S MATH

Use the following information to answer the four questions:

- Purchase unit for frozen hamburger patties: 20-lb case (4-oz patties)
- Order period: 2 weeks (14 days)
- Daily usage rate: 2 cases (28 cases per order period ÷ 14 days)
- Lead time: 3 days
- Number of cases used during lead time: 6 cases (2 cases per day × 3 days)
- Safety level: 8 cases

A. What is the maximum number of cases that should ever be in inventory?

B. What is the order point for hamburger patties?

C. How many cases should be ordered at the order point?

D. How many cases should be ordered if an order is placed when there are 20 cases in inventory?

(Answers: A: 28 cases usage rate + 8 cases safety level = 36 maximum cases; B: 6 cases lead time + 8 cases safety level = 14 cases at order point; C: 28 cases (usage rate) [Proof: 14 cases order point − 6 cases lead time = 8 cases at delivery, which equals the safety level]; D: Step A: 20 cases in storage − 14 cases order point = 6 excess cases; Step B: 28 cases usage rate − 6 excess cases = 22 cases to order [Proof: 20 cases in storage − 6 excess cases = 14 cases at delivery] 14 cases available + 22 cases ordered = 36 maximum cases)

For example, the manager at Pascal's Deli has set a par level of 6 cases of lettuce, at 24 heads per case. She orders lettuce each Friday for Monday delivery, and each Thursday for Friday delivery (*Exhibit 3.8*). To prepare for each order, she counts the number of cases available and "rounds down" to the nearest full case. On Thursday she has 3 full cases and an opened case. Since only full cases are included in her par calculations, she notes 3 cases in inventory.

Exhibit 3.8

The manager knows that she uses about two cases between Friday and Monday (the days between order and delivery), and she also recalls that more lettuce is available in the open case. Therefore, she believes her estimate will not create a stock out.

Using this information, she determines the need to order 5 cases from her local vendor:

6 cases	**–**	**(3**	**–**	**2)**	**=**	**5**
Par level		**Cases available**		**Cases used before delivery**		**Cases to order**

If her usage estimate is correct, the par inventory level will be maintained when the delivery arrives:

3	**–**	**2**	**=**	**1**
Cases available		**Cases used**		**Cases available at delivery**

1	**+**	**5**	**=**	**6**
Cases available at delivery		**Cases delivered**		**Cases at par inventory level**

The manager quickly determined the number of cases of lettuce to order in three steps:

- Remember the par level.
- Determine the number of cases available.
- Subtract the number of cases likely to be used between order placement and delivery from the number of cases available.

Managers decrease par levels if they notice that the quantities on hand are increasing. They also increase par levels if they notice decreased quantities immediately before deliveries. The process just described is a "trial and error" method that, over time, works well in many operations.

SELECTING VENDORS

Managers must make decisions regarding which vendors will be requested to quote prices for the products needed by their operation. Some managers purchase products from whichever vendor offers the "right" price. However, the best vendors also help the manager attain other purchasing goals: to purchase the right products at the right time with the best quality and in the right quantities.

Most managers want vendors with these characteristics:

- They consistently provide the required quality of products and have reasonable prices.
- They meet product delivery schedules, and they provide useful support services including information to help buyers and managers make purchasing and operating decisions.
- They take ownership of problems when they occur and are mutually interested in providing value to customers.
- They have similar values about ethical business relationships and make communication an easy process.
- Hazard Analysis Critical Control Point (HACCP) approved vendors can share liability with an operation if there is a problem with an operation's food.

Determining Supply Sources

Buyers require information about potential vendors to determine which vendors, often from among many, should be approved to receive RFPs from the restaurant or foodservice operation. There are several useful sources for this information:

- **Reputation:** The manager's knowledge and experience with vendors is often the most useful source of information about them.
- **Trade publications:** Effective managers keep up with industry-related information.
- **Electronic marketing information:** There are numerous buyers' guides made available by industry publications, trade associations, chambers of commerce, and other sources.
- **Vendor representatives:** Managers who ask vendors about companies selling noncompeting products may learn about useful supply sources.
- **Trade shows and other professional meetings:** Many meetings include time for attendees to visit exhibits and meet with vendors.
- **Employees:** Employees who have worked at other properties may know about potential products and vendors.
- **Other information sources:** Vendor catalogs, the yellow pages of the local telephone directory, and mailing brochures are examples of additional information sources.

THINK ABOUT IT . . .

A manager's previous experience with a vendor is a factor in determining whether the buyer–seller relationship should continue. How important should experience be in making a purchasing decision?

Some managers conduct on-site inspections of a potential vendor's operation to observe work methods, cleanliness, and general organization. The condition of the vendor's transport equipment is also important.

Range of Vendor Sources

Today's buyers can purchase from many sources. While many managers still use traditional vendors to supply wide product lines and very specific orders, some managers also use wholesale buying clubs, and cooperative purchasing systems to fulfill their operation's needs.

TRADITIONAL VENDORS

Broad line vendors sell a wide range of products, often with few alternatives in each category. **Specialty line vendors** provide a narrow product line, but a deep selection within the line.

A manager might purchase frozen bread dough from a broad line vendor who offers a few varieties acceptable to the majority of users. This vendor may also offer hundreds or even thousands of additional products. By contrast, a specialty frozen food vendor may offer many frozen bread dough choices and, if purchase quantities warranted it, could even produce or obtain special products to meet an operation's exact requirements. This vendor would likely offer few, if any, items outside its specialty line of breads.

There are advantages to purchasing from broad line vendors:

- Reduced unit costs with increased purchase volumes
- Decreased costs for purchasing and accounting aspects of order placement
- Shorter purchasing time because the buyer must interact with only one vendor
- Less time for product receiving activities because only one order will be received
- Reduced time needed to interact with several specialty line vendors

There are also potential disadvantages to relying on broad line vendors. These include the possibility that salespersons may lack detailed product information about all products sold. There might also be a lack of variety if specialized products are required.

As noted, the advantage of using a specialty line vendor is the wider variety of items that will be available for the specific product lines that are sold. Challenges involved with the exclusive use of specialty line vendors include the need to interact with many different supply sources. This involves the additional time and money associated with placing, receiving, and paying for orders.

WHOLESALE BUYING CLUBS

One-stop shopping sources also include wholesale stores such as Sam's Club and Costco in which managers buy products at the vendor's location.

Wholesale buying clubs offer advantages for some operations, especially small ones. Managers may be able to call in their orders, which can then be ready for pickup when they arrive at the store. Charge accounts may also be available. Some buying clubs even deliver products if minimum delivery size requirements are met.

Increasingly, grocery stores offer a variety of products in large purchase unit sizes. In some cases, they provide another reasonable alternative. Local merchants may become the vendor of choice for some operations.

COOPERATIVE (POOL) PURCHASING

A cooperative (pool) purchasing system involves several operations combining orders for products of the same quality. One order is then submitted to the vendor with the lowest quoted price, and products are delivered to each operation.

Some state hospitality associations and for-profit co-ops may offer these services. Participants note lower prices as an advantage.

PRODUCT ORDERING PROCEDURES

The **ordering** process occurs when a buyer makes specific commitments to a vendor for a specific purchase. Managers must do several things as they prepare an order:

- Decide whether it is better to "make or buy"
- Determine what to buy
- Determine the quality to buy
- Decide how much to buy

Managers in small operations may make these decisions by themselves or with help from staff members. Managers in larger operations may make these decisions with assistance from purchasing staff.

Order placement activities may involve only conversations in small properties, while extensive written communication is typically used in larger organizations. Examples include purchase specifications and POs, discussed earlier in this chapter. The basic principles required for effective purchasing are similar regardless of an operation's size. Buyers must always tell vendors exactly what they require.

If there are frequent problems that, from the buyer's viewpoint, are caused by the vendor, the buyer has less interest in placing additional orders. Also, vendors are less likely to provide value-added services for their "problem" accounts.

Pricing Concerns

Vendor prices are an obvious ordering concern. However, managers want the right price, not just the lowest price. Few managers think the best meals are the least expensive, and a focus on the lowest price usually indicates misunderstanding about how prices are determined.

Four primary factors influence the prices buyers pay for products:

- Prices reflect product costs. Vendors must sell their products for more than they pay for them to make a profit and stay in business.
- Prices reflect consumer demand. When products are scarce and many buyers want them, the prices charged will generally be high. In other cases, such as expensive bottled waters, it is the consumers' willingness to pay that most influences price.
- Prices reflect vendor-provided services and information. Receiving a low price on a product that is not delivered when needed is not a bargain. Also, managers know that the value of information they receive from their vendors and are willing to pay for it.
- Prices reflect vendor quality. Vendors with a reputation for quality food and outstanding service can charge more for their products than vendors who are not dependable.

FACTORS AFFECTING PRICING

Managers must decide whether to buy products from one or more vendors. Generally, using more vendors requires additional time for ordering, receiving, and invoice processing. However, a single-source vendor may charge higher prices if the vendor believes there is no competition. Therefore, buyers often split their business among several vendors.

A vendor's cost for delivering a $1,000 order is not much different from the cost for delivering a $100 order. Therefore, when this delivery cost can be spread across more items, a lower per unit price can be charged.

Buyers who do not pay their bills when due typically pay more than their competitors for similar products. Vendors add the extra cost of carrying "slow-pay" accounts to the prices charged these buyers.

Manager's Memo

Buyers and sellers sometimes negotiate pricing and other parts of the purchasing agreement. Negotiation involves the buyer and seller reaching agreement about disputes, determining courses of action, and bargaining for their individual and mutual advantage. Compromise and agreement with an incentive to overcome each other's concerns are part of successful negotiations.

Successful negotiation allows both parties to "win," and this is the desired outcome for a long-term relationship. A vendor granting a price concession might desire a long-term relationship with the manager, and a buyer may pay more if a special delivery time can be arranged.

DISCOUNTS AND REBATES

Many vendors offer discounts, which are deductions from the established price buyers normally pay for products. Discounts are offered to selected buyers based on some characteristic or action. Buyers should ask about all discounts offered by their vendors so they can pay a low **net price**: the total or per-unit amount paid for products after all discounts have been applied.

There are several types of discounts:

- **Prompt payment discount:** These discounts are common because vendors want to reward customers who pay their bills on time.
- **Quantity discount:** Many vendors offer quantity discounts to encourage customers to increase their purchases.
- **Customer status discount:** Vendors may offer status-based discounts depending on their objectives. Factors may include length of customer relationship, nonprofit business status, annual purchase volume, or membership in a specific company, brand, or chain.
- **Special discounts (promotions):** These discounts may be associated with holidays, seasons, or events. They can be offered when a seller wants to clear inventories of slow-moving or discontinued products.

Rebates are discounts offered *after* a purchase has been made at the normal selling price. They are often used to introduce new products or to increase awareness and sales volumes of existing products. *Exhibit 3.9* shows the impact of discounts and rebates on the net price of 1 case of sliced pears containing six #10 cans.

Exhibit 3.9

NET PRICE COMPUTATION FOR ONE CASE OF PEARS

		Case Price (6 #10 cans)
Normal case price		$42.00
Less seller's 2% prompt payment discount	$0.84	
Less manufacturer's after-purchase rebate	$3.00	
Less total discounts and rebates		$ 3.84
	Net price	$38.16

If 6 cans are purchased for $38.16, the cost of 1 can is $6.36:

$38.16	÷	6	=	$6.36
Cost per case		**Cans per case**		**Cost per can**

This per-can price can be compared with different vendors when evaluating competitive costs.

Product Ordering Basics

The ordering process depends on the operation. Small operations that use decentralized purchasing and larger operations with more centralized purchasing systems use different ordering procedures.

SMALL OPERATIONS

Small operations typically use ordering procedures like these:

- Persons responsible for purchasing are identified. In a very small operation, the owner-manager may purchase everything. If a chef, beverage manager, and dining-room manager are employed, they may purchase products needed for their departments (*Exhibit 3.10*).

Exhibit 3.10

- Department heads can spend to a pre-established limit such as $1,000, after which owner or manager approval is required.
- Department heads make basic purchasing decisions including the quantities of products needed and the vendors from whom prices should be requested.
- Department heads complete required purchase documents determined by operation policy and may keep them until orders are delivered. Then they or an employee sign delivery invoices, and the original purchase documents and delivery invoice are sent to those who pay the invoices and record the purchase.

LARGE OPERATIONS

Exhibit 3.11 reviews the basic steps that might be used for ordering in a large operation. Please note that all of these steps can be automated.

Exhibit 3.11

ORDERING IN A LARGE OPERATION

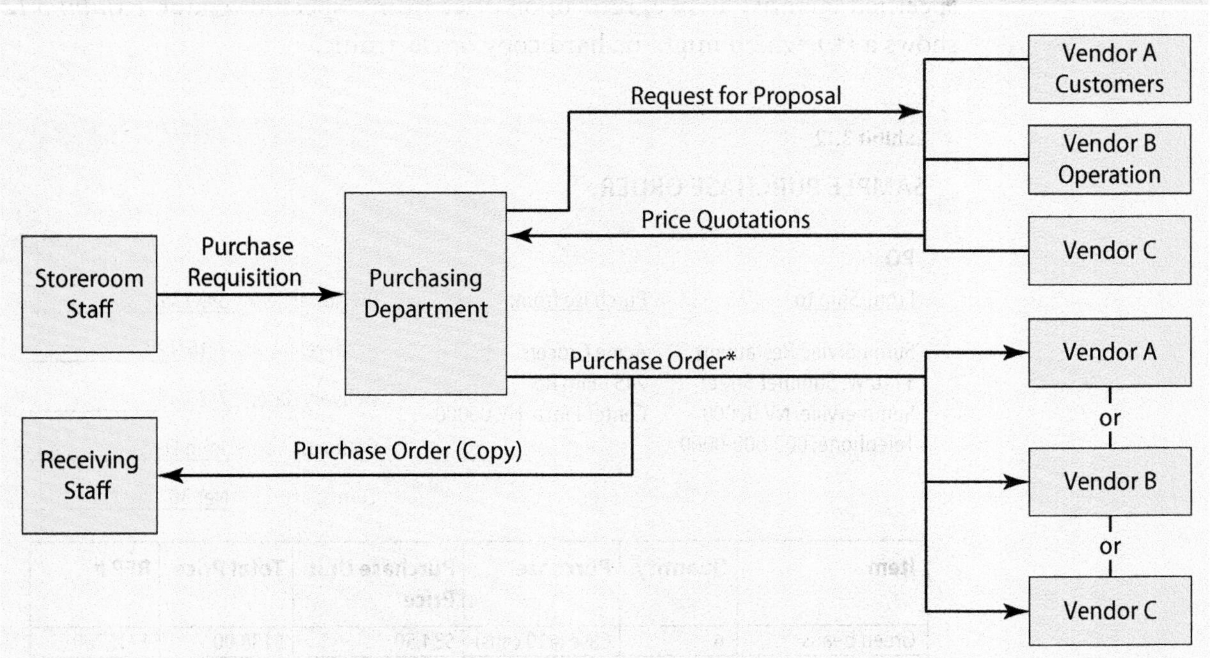

In some restaurant or foodservice operations, buyers send copies of POs to accounting personnel when orders are placed. In other establishments, receiving personnel attach their copy of the PO to the delivery invoices that are then sent to accounting.

The process shown in *Exhibit 3.11* happens as follows:

- Purchasing managers order products when they receive a purchase requisition indicating that additional products are required.
- Managers then issue a request for proposal (RFP) to vendors.
- Vendors interested in providing the products return a price quotation indicating their current selling prices for items meeting the operation's quality requirements, which are found in purchase specifications.
- After analyzing price quotations, a PO is issued to the chosen vendor, and a copy is routed to receiving staff for use when products are delivered. It is important to note that when manual purchasing systems are used, the RFP and the PO are frequently the same document. Prices are requested, the RFP is returned, and a copy is then returned to the vendor as a PO to accept the seller's prices.

Purchasing staff may also route a copy of the PO to accounting staff, who retain and then compare it to the delivery invoice provided by receiving staff after product delivery.

A Close Look at Purchase Orders

POs are used to approve the purchase of products of a required quality in a specified quantity at an agreed-upon price from a specific vendor. *Exhibit 3.12* shows a PO, which might be hard copy or electronic.

Exhibit 3.12

SAMPLE PURCHASE ORDER

PO

From/Ship to:	Purchase from:		
Summerville Restaurant 1710 W. Summer Street Summerville, NV 00000 Telephone: 000-000-0000	Acme Grocers 2451 Elm Rd. Center Place, NV 00000	PO no.	34X135
		PO date:	7/16/12
		Delivery date:	7/20/12
		Contact:	John Davis
		Terms:	Net 30

Item	Quantity	Purchase	Purchase Unit Price	Total Price	RFP #
Green beans	4	Case (#10 cans)	$34.50	$138.00	1715700
Flour, all-purpose	3	Bag (50 lb)	22.17	66.51	1715700
			Total	$783.17	

Terms and Conditions:
This PO expressly limits acceptance to the terms and conditions stated above and included on the following page. Any additional terms and conditions are rejected.

Buyer: John Davis Date: 7/16/12

The buyer listed in *Exhibit 3.12* had requested prices from several vendors and selected Acme Grocers to provide the order. Specifications had previously been sent to all the vendors, and both the buyer and seller know that prices quoted must be based on products meeting these quality requirements.

The PO indicates terms of net 30 in the top right corner. This means the total amount of the invoice ($783.17) must be paid within 30 days of the delivery date. Sometimes buyers negotiate a discounted price for faster payment. For example, payment terms might read 2/10; net 30. This means the seller will offer a 2 percent discount if the total invoice is paid within 10 days; otherwise the amount is due within 30 days.

A copy of the approved PO is typically sent to the vendor and the property's receiving staff. Additional copies may also be retained by the purchasing department and be forwarded to accounting staff.

In most large-volume operations, all orders except those paid for by petty cash are approved with a PO. **Petty cash** is a fund with a limited amount of money that is used to make infrequent and low-cost purchases.

Expediting Procedures

Managers may sometimes expedite deliveries because of problems caused by purchase requisition, or when inventory management errors occur. Weather conditions, strikes, and vehicle loading errors are examples of vendor-related reasons for delivery delays.

Managers should consider the need to expedite purchases as an opportunity to correct problems. They should work with vendors to resolve vendor-related causes of product shortages and quality issues. Problems caused by the manager's organization must be addressed to help ensure the timely delivery of products of the proper quality.

> **OPEN FOR BUSINESS — RESTAURANT TECHNOLOGY**
>
> Software developers offer systems that help merge purchasing, inventory, and menu item sales functions and perform many calculations that managers have traditionally performed manually. These systems provide accurate information, the calculations are done automatically, and time is saved that managers can use for other purposes.
>
> For example, point-of-sale (POS) systems can count the number of each item sold. They can then determine the amount of inventory decreases that should have occurred based on these sales. Managers are alerted when product order points are reached, and ordering decisions can then be made.

PURCHASING FOLLOW-UP

Products that are purchased must be paid for. In very small operations, the owner may be the manager, buyer, and bill payer. In larger operations, managers may do the purchasing, while someone else is responsible for paying bills.

The responsibilities of accountants extend far beyond their role in purchasing. However, since a large percentage of an operation's revenues is used to purchase required products, this task must receive a high priority.

Payment Procedures

"Every penny counts!" is an old saying that is still true today. Vendors must be paid what they are owed. However, they should not be paid more.

IMPORTANCE OF DOCUMENTATION

Preparing delivery invoices or vendors' statements for payment requires communication, information, and coordination between purchasing, receiving, and accounting staff. The basic process is the same regardless of whether a manual or computerized system is used. Those with accounting responsibilities require documentation to prepare payments in varying sizes of operations:

- **Small operation with owner-manager present:** Products are ordered and may be received by the manager. A copy of the PO or vendor's delivery invoice is used by the owner-manager for payment purposes. The owner may pay the bill or transfer documentation to an off-site bookkeeper for payment.
- **Small operation with absentee owner:** In this situation, the manager of the property or employees will be responsible for purchasing and receiving, and the PO or delivery invoice may be routed to the bookkeeper for payment. The vendor may send a separate copy of each delivery invoice directly to the absentee owner or bookkeeper so both copies of the delivery invoice can be matched.
- **Large operation with separate purchasing and accounting departments:** A copy of the PO is sent from purchasing to receiving staff, who check incoming orders against it. The PO and signed delivery invoice are then routed to the purchasing department to be matched before being sent to the accounting department as payment authorization. Sometimes POs and delivery invoices are first sent to the chef or operation director for review before they are routed to purchasing staff.

PAYMENT METHODS

Two basic methods can be used to pay vendors:

- **Pay by invoice:** Approved delivery invoices and other supportive information are manually or electronically filed for payment on a specific date, and these bills are paid when due. If an approved invoice must be paid by July 16, it may be "pulled" on July 10 for final review, signature, and mailed on July 11 to allow time for mail delivery.
- **Pay by statement:** Processed documentation including delivery invoices is filed by vendor until the vendor sends a statement of account. For example, a vendor who delivers bread daily may request payment every two weeks. This vendor sends a statement to the operation every two weeks listing invoice information for the bread deliveries received during that time. Then all invoices covered by the statement, less adjustments, if any, are paid at the same time.

Exhibit 3.13 shows a sample vendor statement.

Manager's Memo

Delivery invoices indicate the products and quantities of each product that were delivered and for which vendors expect payment. These documents must be carefully studied to ensure there are no quantity or price differences between items on the PO and those listed in the delivery invoice. Accounting staff must confirm the correctness of all extensions (calculations such as the number of items purchased multiplied by the price of each item). They must also file the invoice with the applicable documentation and pay it at the proper time.

Exhibit 3.13

SAMPLE VENDOR STATEMENT

Statement 107643

Bayside Produce
117 Bayside Street
Anytown, Any State 00000
Telephone: 000-000-0000
Fax: 111-111-1111

Account no. 1735210
Delivered to: Garden Inn Restaurant
300 Garden Lane
Anytown, Any State 00000
Attention: Jack David, Director of Purchasing
Telephone: 000-000-0000

Invoice No.	Delivery Date	Amount Due	Adjustment	Net Amount Due
10711	2/10/12	$173.59		$173.59
10928	2/13/12	310.80	($21.55) Credit memo #2138	289.25
12541	2/18/12	190.51		190.51
13401	2/23/12	290.18		290.18
			Total	$943.53

Payment due upon receipt. Please send payment to the above address.

Duplicate: Please return top copy with payment, and retain second copy.

Thank you.

Other Purchasing Issues

Attention to invoicing represents a significant amount of a manager's purchasing focus. However, managers must also pay special attention to the management of credit memos and petty cash funds.

CREDIT MEMOS

Accounting staff may have some need to interact with vendors about pricing and related concerns, even with proper documentation. One issue that arises as bills are prepared for payment relates to credit memos that identify problems noticed during delivery:

- **Incorrect price charged:** The agreed-upon price noted on the PO is less than that on the delivery invoice.
- **Backorder:** The product is included on the delivery invoice but was not received.

Exhibit 3.14

- **Short weight or count:** More products were ordered and included on the delivery invoice than were delivered.
- **Items rejected because of unacceptable quality:** Items do not meet purchase specification requirements and are rejected on attempted delivery (*Exhibit 3.14*).

Credit memos are typically issued by the vendor at the time of delivery. They should be treated like cash because they represent a reduction of the amount owed based on the delivery invoice. Most frequently, they are attached to the applicable delivery invoice for routing to purchasing employees and then on to accounting staff.

PETTY CASH PURCHASES

Managers typically minimize the use of cash payments for purchases because checks allow for better documentation and control of expenses. However, minor expenses such as office supplies and even "emergency" food purchases from grocery stores are sometimes best paid from a petty cash fund because it is more practical and less expensive than issuing a check.

The petty cash bank must be kept locked in a secure location such as the manager's office, and one person should be responsible for it. There are several accepted rules for petty cash funds:

- The amount of money in the petty cash fund should be based on the normal value of petty cash purchases for a specific time period.
- A check payable to "petty cash" should be written to establish the initial petty cash fund.
- When purchases are needed, cash is removed from the cash bank and a petty cash voucher for the amount of money removed is placed in the bank. A **petty cash voucher** is a slip signed by the person responsible for the petty cash fund that authorizes the withdrawal of cash from the fund for a purchase.
- The change from the transaction and the purchase receipt are returned to the manager, who confirms that the amount (Cash change + Receipt amount) equals the amount removed.
- When more money must be put into the petty cash fund, another check is written payable to "petty cash" for the value of all paid receipts in the bank. This check, converted to cash, replenishes the fund to its original cash amount.
- At any point, the money available in the petty cash fund plus the amount of all documented receipts should equal the original amount of money allowed for petty cash (Cash amount + Purchase receipts = Total petty cash fund).

SUMMARY

1. **State the objectives of an effective purchasing process.**

   Purchasing has five basic objectives: to purchase the right product from the right source at the right price in the right amount, and at the right time. These objectives drive purchasing activities in large operations with centralized purchasing systems and smaller operations using decentralized purchasing systems.

   Managers are responsible for many tasks including identifying and selecting vendors, negotiating prices, ensuring that products of the proper quality and the correct quantities are ordered. Follow-up activities include managing purchase documents and facilitating payment of delivery invoices.

2. **Define quality, and review purchasing procedures that help ensure that operations obtain products of the proper quality.**

   Product quality relates to intended use. Managers must determine what products will be used for and then select a product that is suitable for that purpose.

   Purchase specifications describe quality requirements. They should be given to vendors with the instructions that the prices they quote should be for products meeting these requirements.

   Specifications should be developed with input from the operation's employees and vendors, who can provide suggestions and ensure that the operation does not limit the number of vendors who can provide products.

3. **Describe basic procedures for determining the quantity of products to purchase.**

   Buying too much ties up capital, impacts cash flow, and requires additional storage space. It also increases the risk of theft and quality deterioration. Buying too little can create stock outs that impact production and, potentially, disappoint customers.

   Primary factors that impact purchase quantity decisions involve estimated production volumes, discounts, storage space, and cash flow. Order quantities are also influenced by minimum orders established by vendors, anticipated price changes, and discounts and rebates offered.

   Many managers determine quantities to order using a minimum–maximum inventory system. This system indicates the minimum quantity below which inventory levels should not fall and the maximum quantity above which levels should not rise. Purchase quantities for perishable products are often determined using a simple par inventory system. This system involves determining the maximum amount of products that should be available based on product usage. Then an amount is ordered that will bring the quantity available to the maximum amount when the products are delivered.

4. **Identify concerns important when selecting product vendors.**

   Many concerns other than lowest price are important when selecting vendors. Those selected should consistently provide the required product quality, meet delivery schedules, and provide support and information. Information about potential vendors can be learned through reputation, trade publications and electronic marketing, vendors of noncompeting products, trade shows and other meetings, and employee suggestions.

In addition to traditional broad line and specialty line vendors, some managers purchase from wholesale buying clubs or participate in cooperative (pool) purchasing.

5. **Explain basic product ordering procedures.**

   Vendors' prices are an important concern in product ordering decisions. The prices buyers pay are influenced by product cost, consumer demand, service and information provided, and quality. Some vendors offer discounts, which are deductions to established prices, and rebates, which are deductions that are taken after payment.

   Small operations typically allow department heads to order products based on the operation's policies. Larger-volume establishments use a more centralized system. Purchase requisitions are sent from each department to a manager who sends a request for price quotation (RFP) to approved vendors. Those interested respond, and a PO is used to summarize the purchase agreement.

6. **Describe basic payment methods and the management of credit memos and petty cash funds.**

   Payment documentation is important. Managers can pay by invoice where payment is made separately for each delivery invoice, or they can pay based on a vendor's statement of account that summarizes several invoices.

   Vendors issue credit memos to reflect pricing differences created when the amount of product reported on the delivery invoice differs from the amount actually received. Problems can occur when products are rejected for quality or other reasons and when incorrect quantities are delivered. Managers must handle and process credit memos carefully because they represent reductions in the amount due to vendors.

   Many properties use petty cash funds to purchase inexpensive items. A cash advance system should be used in which money is taken from the petty cash fund and then, after purchase, the purchase receipt and change are replaced in the fund. The value of the petty cash fund should always be the amount of current cash plus paid-out vouchers, which should equal the approved petty cash fund amount.

APPLICATION EXERCISE

Your menu planning team has added a new Cajun and Creole menu item, and it requires Andouille sausage. One of your approved meat vendors says he can obtain it fresh for you but a minimum order, specified in a specific number of pounds, must be purchased at any given time. It is also available frozen in smaller quantities. Although not preferred, frozen sausage does provide an alternative because a large order can be stored without quality problems arising. You are also aware that there is at least one specialty line vendor in the area that can sell a fresh product in smaller amounts for a higher price.

1. What must you think about and what procedures should you use to determine the quality of sausage that you should buy?
2. What procedures should you use to determine from which vendor you should purchase the product?

REVIEW YOUR LEARNING

Select the best answer for each question.

1. **Which document summarizes the purchase agreement?**
   A. Purchase order
   B. Purchase requisition
   C. Purchase specification
   D. Request for price quotation

2. **A centralized purchasing system is one with**
   A. specialized staff.
   B. a vendor handbook.
   C. expediting methods.
   D. competitive bidding.

3. **Which factor most affects an establishment's product quality requirements?**
   A. A product's purchase cost
   B. A product's planned use
   C. Industry food cost data
   D. Industry specifications

4. **Which document describes the quality requirements for products that are ordered?**
   A. Standard recipe
   B. Delivery invoice
   C. Vendor statement
   D. Purchase specification

5. **Which is the purpose of a purchase requisition?**
   A. It tells buyers that more products must be purchased.
   B. It is used to expedite receipt of undelivered products.
   C. It informs vendors that buyers want to order products.
   D. It asks vendors for current prices for needed products.

6. **The value of the petty cash fund should always be the amount of**
   A. current cash minus paid-out vouchers.
   B. current cash plus value of petty cash check.
   C. current cash plus paid-out vouchers.
   D. current cash minus value of petty cash check.

7. **In a minimum–maximum inventory system, which factors determine the order point?**
   A. Lead time + Usage rate
   B. Safety level + Usage rate
   C. Lead time + Safety level
   D. Safety level + Minimum delivery

8. **A par level has been set at 8 cases of frozen French fries. The manager has 4 full cases and 1 partial case in inventory. She will use 2 cases between order and delivery. How many cases should the manager order?**
   A. 4
   B. 5
   C. 6
   D. 7

9. **Which is the primary advantage of purchasing products from broad line vendors?**
   A. Longer time payment plans
   B. Greater variety within product lines
   C. Less time spent in purchasing activities
   D. Better assurance of high-quality products

10. **What is a rebate?**
   A. A discount given to buyers for large quantity purchases
   B. A deduction from selling price after the price is set
   C. A discount offered after a purchase has been made
   D. A deduction given for pickup instead of delivery

4 Product Receiving, Storing, and Issuing

INSIDE THIS CHAPTER

- Product Receiving Procedures
- Product Storage Procedures
- Product Issuing Procedures
- Technology and Receiving, Storing, and Issuing

CHAPTER LEARNING OBJECTIVES

After completing this chapter, you should be able to:

- Explain each of the steps in an effective product receiving process.
- State procedures for maintaining product quality, reducing the possibility of stock outs, and controlling costs during storage.
- Describe procedures for controlling product costs when products are issued from storage to production areas.
- Explain ways that technology can assist with receiving, storing, and issuing tasks.

KEY TERMS

backorder, p. 112
broken case storage, p. 113
cost of goods sold (COGS), p. 116
credit memo, p. 105
first in, first out (FIFO), p. 106
income statement, p. 118
inventory turnover rate, p. 116
issue requisition, p. 121
marking (product), p. 106
operating budget, p. 118
package inspection program (employee), p. 120
receiving report, p. 107
reduced oxygen packaged (ROP), p. 109
slack-out seafood, p. 107
transfer, p. 119
workflow, p. 114

CASE STUDY

"I like the old way better, don't you?" Armando asked Elliot, another cook at Broadway Diner.

"If you mean the big deal now about issuing and storing, yes!" replied Elliot.

"I've been working here for two years. I came to work last week and saw the new shelving unit like a cage in our walk-in refrigerator. It sure takes a lot of time to fill out the new form when we want items locked that are in that cage," said Armando.

"Also," he continued, "sometimes I don't think about everything I need for the next couple of hours, and then I've got to go pester the boss. That bothers him and wastes my time."

"The boss said this system helps with checking things against the budget and determining whether we make a profit," said Elliot. "I think things have been going well and there is no reason to change. I really think the boss just doesn't trust us and wants to make things harder for us!"

1. How can use of a lockable "cage" in a walk-in refrigerator help control product costs?
2. What should the manager have done before purchasing the cage, and what should he or she do now?

PRODUCT RECEIVING PROCEDURES

Product receiving involves a careful check of incoming products before they are placed in the proper storage areas. Plans and procedures must be made, implemented, and consistently followed to effectively control products when they are received. Without these measures, product quality may not be acceptable, and costs may be higher than they should be.

Receiving is more than allowing the employee closest to the back door to sign the paperwork for deliveries and move the products into storage. Basic receiving practices are needed, and special concerns must be addressed as the process is developed.

Getting Ready for Receiving

Effective product receiving requires careful attention to employees with receiving responsibilities, the receiving area, and the tools and equipment used.

RECEIVING STAFF

Trained staff members should perform receiving tasks, and the widely used policy that almost anyone can sign the delivery invoice should be avoided. A delivery invoice is a document signed by an employee of the operation when products are delivered to transfer ownership from the vendor to the establishment.

Exhibit 4.1

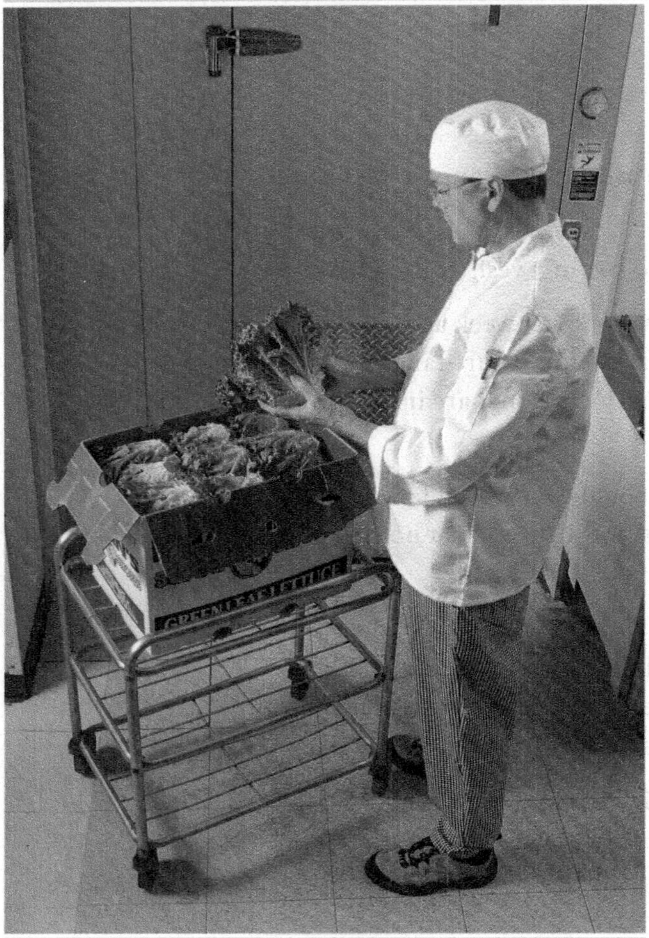

Some receiving procedures, such as checking the quantity of incoming products against the delivery invoice, are clerical and require limited training. However, the ability to recognize product quality and ensure that it meets the property's standards requires extensive training and experience. This is especially critical for fresh produce, meat, and seafood that are not typically purchased by a brand name that can be double-checked on a package label (*Exhibit 4.1*).

Careful inspection of these items is important to verify that purchase specification requirements are met. Recall from chapter 3 that purchase specifications describe the quality requirements that must be met by products purchased for the operation. Training programs for receiving positions should include shoulder-to-shoulder interactions with experienced buyers, the food-production manager, or others who can help trainees learn how to confirm that incoming products meet quality requirements.

Large-volume operations may have receiving and storeroom clerks with full-time responsibility for receiving and storing activities. Employees such as the production manager or food-production team members may assume these duties in smaller-volume properties. However, the basic practices for managing food products effectively apply to all operations, and they must be performed by a trained employee regardless of whether the activities are his or her only responsibility, or one of many.

Successful receiving staff have several abilities in common:

- They maintain food safety standards. Food safety concerns are important aspects of quality, and they should be considered in the property's purchase specifications and receiving procedures.
- They can use appropriate technology. Technology can expedite receiving and storage procedures and maintain accurate information. It is increasingly used by large-volume operations and will likely be more widely adopted in smaller-sized properties in the future.
- They can perform physical requirements of the job. Cases of products can weigh 30 pounds or more. Flour and sugar are available in 50- or even 100-pound bags, and these may be purchased because of lower purchase unit prices. A purchase unit is the weight, volume, or container size in which a product is purchased. For example, chicken may be purchased by the pound, and tomato ketchup can be purchased by the gallon or case of six #10 cans.
- They are able to resolve problems. Issues such as what to do if incoming products do not meet quality requirements or improper quantities are delivered will sometimes arise. Trained receiving staff can deal with these concerns in a way that protects the interests of the operation.
- They have an attitude of concern. Concerned receiving staff are committed to helping the operation move toward attaining its mission, and they recognize their role in helping serve customers.

RECEIVING AREA

The best location for a receiving area is close to the kitchen's back door. However, in very large organizations it is often part of or close to a loading dock located away from food storage and production locations. Small-volume properties with limited space may have little more than a hallway or entry area.

Adequate space to assemble all incoming products is necessary for counting and weighing products, and a receiving scale is needed. Space to locate transport equipment is likely required, a lot of space if large quantities of products are purchased.

Manager's Memo

The Americans with Disabilities Act (ADA) is a federal law that prohibits discrimination against people with disabilities in employment, and it applies to private employers including restaurant or foodservice operations with 15 or more workers. The ability to lift heavy or bulky containers may or may not be a Bona fide occupational qualifications (BFOQs): a qualification judged reasonably necessary to safely or adequately perform all job tasks in a position. Managers must know about and consider how to modify receiving tasks to make jobs available to otherwise qualified persons.

RECEIVING EQUIPMENT

Useful receiving tools include pocket thermometers, a clipboard, and plastic tote boxes or other containers to transport ice for products such as fresh poultry and seafood. Increasingly, personal digital assistants (PDAs), notebook or laptop computers, and other wireless devices are used to access purchasing and inventory records required while receiving. A platform scale is required, and a desk or a filing cabinet is sometimes useful. A two-wheeled dolly, cart, or other equipment to transport incoming products to the proper storage areas is helpful in large-volume operations that receive large quantities of products.

Steps in Product Receiving

There are four basic steps required for effective receiving. *Exhibit 4.2* reviews these steps.

Exhibit 4.2

STEPS FOR EFFECTIVE RECEIVING

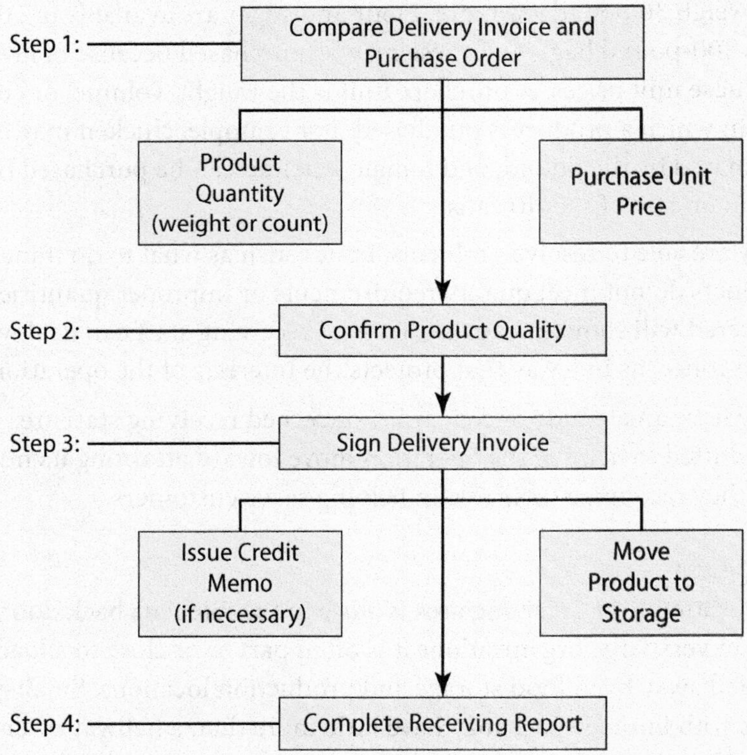

COMPARE DELIVERY INVOICE AND PURCHASE ORDER

In the first step, the delivery invoice should be compared to the copy of the purchase order. A purchase order is a document used to request prices from vendors or to inform them about the products to be purchased. Comparing the delivery invoice with the purchase order helps ensure that the quantity and price of products agreed upon when purchased are correct when they are delivered. Items such as cases of canned goods can be counted, while items such as fresh meat ordered by the pound must be weighed.

CONFIRM PRODUCT QUALITY

Exhibit 4.2 indicates that the second step in the receiving process is to confirm that the quality of the incoming products meets purchase specification requirements. This step is probably the most challenging because purchase specification standards must be easily observable. If receiving staff have concerns about product quality, the production manager or purchasing agent should be asked for assistance.

Products that do not meet the property's quality and quantity standards must be rejected. Accepting lower-quality products wastes the operation's money and hinders its ability to meet the quality standards required by customers.

Several steps should be taken when unacceptable food items have been identified:

- Reject the products if there are obvious damage, safety, or quality concerns and receive a credit memo from the supplier's delivery person. A **credit memo** is a document used by accounting staff to adjust information about product quantities and costs included on a delivery invoice. Its purpose is to ensure that the operation pays only for the actual products that are acceptable and have been received.
- Notify the vendor about the situation. He or she should immediately send a substitute for the rejected products.
- Note the rejected products on the purchase order and delivery invoice.
- Notify the operation or food-production manager according to policy.
- Inform the appropriate manager when the problem is resolved.

The credit memo should be cosigned by the receiving employee and the vendor's delivery person. This is important because the signed delivery invoice is the basis for the payment due to the vendor.

SIGN DELIVERY INVOICE

If no variations from product standards or other quality-related problems are observed, the delivery invoice can be signed (step 3 in *Exhibit 4.2*).

After the delivery invoice has been signed, receiving staff must quickly move products to their proper storage areas. There are three reasons to do this:

- To help prevent loss of quality in refrigerated or frozen products
- To best ensure that no food safety problems will arise
- To reduce the possibility of product theft

THINK ABOUT IT . . .

Sometimes delivery persons want to rush receiving and correct errors later. When this happens, many managers inform vendor sales representatives.

What would your policy be if delivery staff tried to rush your receiving staff?

Receiving staff members are typically responsible for placing incoming products into storage. This involves several activities:

- Writing the date of receipt and the purchase unit price on the incoming products (*Exhibit 4.3*). This task is called **marking**, and it is used to help ensure that the oldest products are issued first. Marking also helps with calculating the cost of products in inventory in some systems.
- Rotating the older stock by placing all incoming products behind or under those items already in inventory. Managers want to minimize the time that products are in storage. To do this, they establish a basic stock rotation program. One popular example is **first in, first out (FIFO)**. As the term implies, products in storage the longest should be the first issued. A simple way to implement this plan is to require that all incoming products be placed behind or under those products already in storage. This makes it easy for production staff to remove the oldest products first. The FIFO system is also made easier when the delivery date is marked on the shipping container. Product marking is discussed later in this chapter.
- Updating perpetual inventory records to increase the balances of products on hand by the quantities received.

Exhibit 4.3

COMPLETE RECEIVING REPORT

A final step in product receiving is to complete a **receiving report**, which is used in operations that calculate food costs on a daily basis. The report separates delivery invoice information used for daily food costing.

At the end of the day or other time period required by the operation's policy, the copy of all purchase orders given to receiving staff and the applicable delivery invoices should be routed to the accounting department. In some operations, the food-production manager or unit manager first reviews these documents. Credit memos should also be attached to documents applicable to specific suppliers.

In some operations, receiving staff verify the invoice extensions to confirm that the purchase unit prices multiplied by the number of purchase units equals the amount on the invoice. For example, if 12 cases of hamburger buns are received and the purchase order indicates that they cost $14.17 each, the extension check should verify a delivery invoice charge of $170.04.

12 cases × $14.17 each = $170.04

Receiving staff may also verify other information to confirm that the total invoice amount is correct. This "first-round" invoice verification will then be followed by additional arithmetic verification performed by the purchasing staff.

It is clear that some time is required for effective receiving. That is one reason many properties have policies against accepting products during especially busy times such as the midday meal period. The actual schedule for incoming deliveries will be established when orders are placed, and deliveries are affected by the forecasted volume of business. For example, there may be relatively few and smaller deliveries during midweek and larger and more deliveries at the end of the week to allow for higher volumes during the weekend.

Special Receiving Concerns

Receiving procedures are impacted by concerns including those relating to quality, food safety, security, and credit memos.

ENSURING QUALITY

It does little good to develop purchase specifications unless they are used to confirm that incoming products meet quality standards. An exhaustive list of ways to identify quality is difficult to develop and depends on the specific items being purchased.

Effective receiving staff can help determine whether incoming products meet quality requirements. For example, sometimes this involves noticing **slack-out seafood** (seafood that was frozen and then thawed to appear fresh so it could be sold at a higher price). At other times, receiving employees may recognize

Manager's Memo

Experienced receiving employees know that it does little good for purchasers to state necessary product quality requirements unless these requirements are confirmed when products are received. They also understand that their operation is very likely to pay for the quality of products ordered, even if that quality is not received.

It is much better for receiving staff to observe that quality-related problems exist before products are accepted than for production employees to note the problems after food items have been issued to the kitchen. Receiving staff have the very important task of ensuring that their operation receives the right quality of products at the agreed-upon price.

that the lengthy and sometimes confusing names of imported wines do not match those on the purchase order. They can check expiration dates on applicable containers and know that produce in the center of a shipping container may not be of the same quality as products on the top layers.

ENSURING FOOD SAFETY

Food safety concerns are very important at the time of product receiving. Employees with responsibility for this task must be trained to check products for improper temperatures, expired code dates, signs of thawing and refreezing, and pest damage, among other concerns.

Clean receiving equipment including hand trucks, carts, dollies, and containers should be available in the receiving area. If products must be washed or broken down and rewrapped before storage, work space as close to the receiving area as possible is important. This prevents dirt and pests from being brought into storage areas or the kitchen.

Products should be inspected immediately on arrival. Receiving staff should check for damaged products and look for items that might have been repacked or mishandled. Spot-checking temperatures of all refrigerated food is also a good practice. Products requiring refrigeration should be put into storage areas as soon as possible.

Specific guidelines for checking temperatures must be followed:

- Insert the thermometer stem or probe into the thickest part (usually the center) of meat, poultry, and fish (*Exhibit 4.4*).

Exhibit 4.4

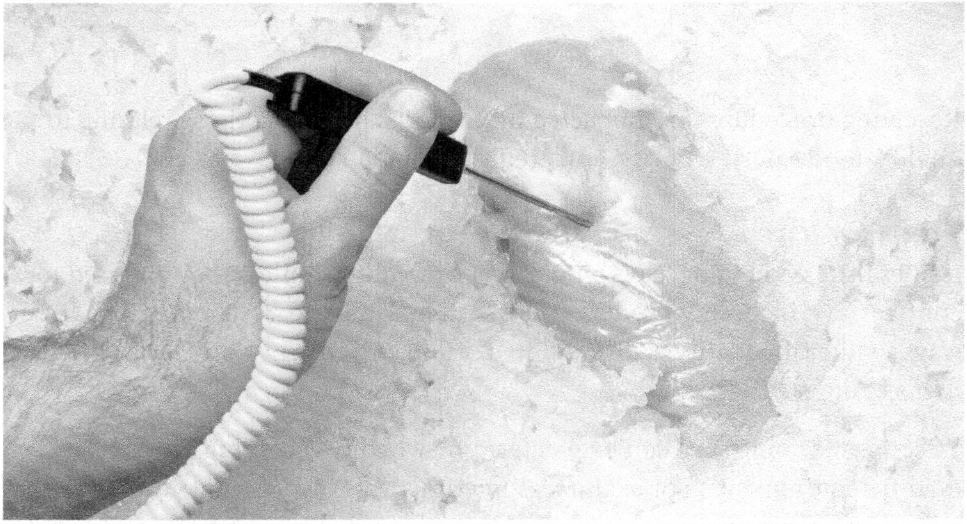

- For reduced oxygen packaged (ROP) bulk food, insert the thermometer stem or probe between two packages. ROP food is contained in a package in which oxygen (a) has been removed, (b) has been displaced with another gas or combination of gases, or (c) something else has been done to reduce the oxygen content to a level below that which is normally found in air. It may also be possible to check product temperature by folding the packaging of refrigerated products around the thermometer stem or probe, but be careful not to puncture the packaging.
- For other packaged food, open the package and insert the thermometer stem or probe into the product. The sensing area must be fully immersed into the product, and the stem or probe must not touch the package.
- For live molluscan shellfish, insert the thermometer stem or probe into the middle of the carton or case between the shellfish for an air temperature reading. Check the temperature of shucked shellfish by inserting the stem or probe into the container until the sensing area is immersed.
- For eggs, check the air temperature of the delivery truck as well as the truck's temperature chart recorder for extreme temperature fluctuations. These, along with high humidity and warm temperatures, may result in the growth of harmful microorganisms.

For specific information about the inspection of incoming products, see chapter 6 of the National Restaurant Association's *ServSafe Coursebook®*.

IMPLEMENTING SECURITY CONCERNS

Security concerns at the time of receiving can cause the operation to pay for products it does not receive. There are several ways this could occur:

- **Wrong weights or counts:** Sometimes small quantities of different products are placed in one box for transport to the buyer. For example, a 20-pound package of frozen steaks may be shipped in a box along with 25 pounds of frozen ground beef. While this box will weigh only slightly more than 45 pounds (20 pounds + 25 pounds + Weight of box), each package should be weighed separately because the costs for steaks and ground beef are different.

  Container weight becomes more significant when fresh seafood or poultry is received in shaved ice inside heavy waxed cardboard containers. When incoming products are packed this way, they should be removed from the containers to determine the weight of the products actually received. Then they can be replaced in the shipping cartons. More ice may need to be placed on the items before they are moved into storage.

- **Missing items:** If 5 cases of frozen green beans appear on the invoice but only 4 cases are on the delivery truck, the delivery person may ask receiving staff to sign the invoice and expect another case next time in an effort to save the trouble of creating a credit memo. This should never be done. First, the same receiving person may not be on duty or may forget about the missing item during the next delivery. Second, a delivery person is unlikely to have the authority to request that products not listed on delivery invoices be placed on the truck.
- **Items of improper quality:** The operation will pay the cost of the product that is on the delivery invoice, so quality substitution can be a serious problem. This occurs when the quality of product described in the purchase specification is not the quality of product delivered. A brand of product ordered may be replaced with another brand, and this problem may not be noticed until production employees see the product issued. Unfortunately, this is likely to be long after the delivery invoice has been signed.

Delivery employees should not have access to nonpublic areas of the operation. Managers know that the best product control occurs when the minimum number of persons can enter storage areas. It is ideal when persons who are not employees are not able to enter these areas.

Hotel food and beverage operations probably have the greatest concerns when this occurs. Consider the problems, including legal liability, that can arise when delivery persons who are not employees gain access to back-of-the-house corridors and employee-only elevators. They can then access other storage areas and even guest rooms.

MANAGING CREDIT MEMOS

A credit memo should be issued by the vendor's delivery person when there is a difference such as weight or count between products included on the delivery invoice and those actually received.

Exhibit 4.5 shows a sample credit memo. The credit memo is typically offered by the delivery person, but it is also wise to have a copy developed by the operation available in case the delivery person does not have one.

Credit memos indicate the amount recorded on the delivery invoice that does not need to be paid because of a problem noted when products are delivered. Therefore, it must be handled in the same way as money. Credit memos are typically attached to the affected delivery invoice by the employee receiving the products. They are then routed with others to promptly reach the manager, purchaser, accountant, or other authorized employee.

Exhibit 4.5

SAMPLE CREDIT MEMO

Date: ______________________ Credit memo no.: ______________________

Vendor: ______________________

____

____

____

Issued to: ______________________

____

____

Account no.: ______________________

For invoice no.: ______________________

Item	Purchase Unit	Number of Purchase Units	Price per Purchase Unit	Total Price
			Total	$

Reason for credit (check):

☐ Backorder ☐ Incorrect quality ☐ Incorrect item

☐ Short count/weight ☐ Incorrect price ☐ Not ordered

Other:

____

____

Authorized signatures:

______________________ ______________________

Vendor's representative Purchaser's representative

RESOLVING RECEIVING CHALLENGES

Receiving and user staff may be aware of problems that are not shared with buyers, and then these concerns cannot be addressed. There are numerous examples:

- **Vendor backorders:** When vendors have problems with their own sources of supply, these products will be on backorder (an order that a vendor cannot fill or ship immediately). There could be a problem that affects all vendors, or a single vendor might have financial problems that affect its ability to purchase for resale. Either way, the buyer should know about these problems.
- **Problems with quantities or PU sizes:** Different packaging sizes may be substituted for those ordered. For example, an order for one 50-pound bag of flour is met with five 10-pound bags. While 50 pounds of flour are received either way, the price per pound is often less for a larger purchase unit. Even with a price allowance, production employees must spend additional time opening bags when filling flour bins (*Exhibit 4.6*).

Exhibit 4.6

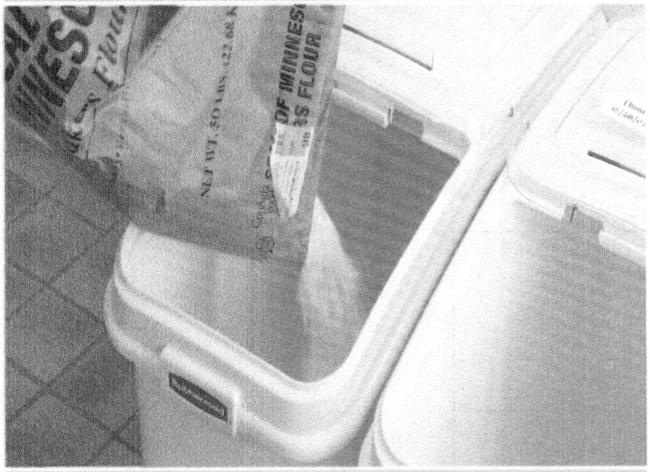

- **Problems with on-site shortages:** Purchasers must always be careful when determining quantities to order to ensure that stock outs do not occur. Problems including product theft, spoilage, errors in inventory calculations, and inaccurate production forecasts can create the need to order products before purchasers normally expect to do so.
- **On-site storage capacities:** Storage space is expensive, and kitchen designers try to plan ideal storage requirements. Unfortunately, "ideal" capacities for an operation planned years ago may be significantly different than for the current operation. Purchasers for high-volume operations with inadequate storage space may have to purchase more frequently with additional costs incurred. Alternatively, they may need to create additional storage space, most often converting other space.
- **Coordination between receiving staff and purchasers:** Those who purchase should "manage by walking around" and be physically present on random occasions when products are received. They can observe products in storage and production areas and have conversations and meetings to learn about products being ordered and challenges when receiving them.

PRODUCT STORAGE PROCEDURES

Incoming products must be stored, and numerous inventory management practices then become important. These are planned and implemented to maintain product quality, reduce the possibility of stock outs, and control inventory costs.

Storage Overview

Experienced managers know that the quality of most food products will never be better than when they are received; products will not improve in storage. The reverse is also true: Product quality can be lowered if proper storage practices are not used consistently.

The operation's financial goals are also impacted by storage practices. If products are stored correctly, all products purchased can be used to generate revenue. However, if products are not properly stored, they must be thrown out and food costs will increase. More food items will have to be purchased at additional cost to generate the planned amount of revenue.

Product quality and cost concerns must be effectively addressed when storage procedures are planned. If this is not done, efforts to maintain quality and cost standards in purchasing and receiving are wasted. Fortunately, the best storage procedures do not require excessive time or costs to implement and maintain.

There are three basic types of food storage areas:

- **Dry storage (50°F–70°F [10°C–21°C]):** This temperature range is best for grocery items such as canned goods, condiments, bakery products like flour and sugar, and herbs and spices.
- **Refrigerated storage (below 41°F [5°C]):** This temperature is best for items such as fresh meat, produce, seafood, and dairy products.
- **Frozen storage:** The proper frozen-storage temperature varies from product to product. A temperature that is good for one product may affect the quality of another product. Frozen storage is for items such as frozen meat, seafood, French fries, and vegetables purchased in this market form.

Some operations use **broken case storage**: an area used to store opened purchase units of products that have been issued. Some operations issue products such as canned goods in full purchase units (cases) even if only partial units (several cans) are needed. If two cans of vegetables are removed from a case containing six cans, the broken case has four cans remaining that are stored in a broken case area for quick access. Some operations also have storage space in work stations including food-preparation and serving areas.

Manager's Memo

Restaurant and foodservice managers may encounter vendors whose delivery staff are authorized to make only "inside back door" deliveries. In other words, delivery persons can move products only from their vehicle to just inside the building. This is good from a security point of view; vendor staff should not be allowed free access to back-of-the-house areas. On the other hand, this can create a problem when receiving or storage areas are located far from production.

Another storeroom location challenge arises when a large facility such as a hotel has several dining locations or kitchens. Then the within-property transport of food products to storage areas or within production areas creates time and control issues that must be effectively addressed.

The location of storage areas is important. The ideal space is close to the receiving area, between the receiving and food-production areas. This arrangement has at least one possible disadvantage: It may be easier for employees to steal items when they exit the operation through the employee entrance or exit (often the back door) if proper controls are not in place.

Storage areas are sometimes affected by remodeling projects, which can occur several times during the life of the building housing the operation. Frozen-storage space may be relocated to an outside area to convert interior space for other uses. Storage space may be moved to remote locations including different levels of the building. Design changes must be made with the understanding that storage space impacts the operation's **workflow** (movement of products through work stations). Storage areas should not just be located "wherever there is room."

Storage and Inventory Basics

Managers should understand four basic aspects of storage and inventory management: quality concerns, physical storage locations, record-keeping requirements, and security concerns.

QUALITY CONCERNS

Products can decline in quality even under very good environmental conditions. For example, suggested storage temperatures might be exceeded. The recommended temperatures noted earlier for different storage areas should be maintained, and they should be monitored with appropriate equipment that is routinely checked for accurate temperature readings.

Storage areas should be cleaned according to a planned schedule, not just when someone "gets around to it." Shelving units should keep products off the floor and away from the walls and ceilings to help with cleaning and to improve air circulation. Managers who maintain the proper storage environment, minimize the time products are in storage, and regularly monitor storage areas are taking important steps to control product quality during storage.

Several practices are very important for handling products in refrigerated storage areas:

- Monitor the temperature of each refrigerator unit at least once during each shift. The internal temperature of refrigerated food should be 41°F (5°C) or lower.
- Schedule regular preventive maintenance for refrigerators to keep them in peak operating condition. This maintenance helps minimize problems that can lead to downtime and the need to throw out food because of unsafe holding conditions.
- Do not overload refrigerators.

- Keep refrigerator doors closed as much as possible.
- Wrap or cover all food properly.
- Store raw meat, poultry, and seafood separately from ready-to-eat food.

General guidelines are also important for products stored in freezers:

- Set freezers to a temperature that will keep products frozen.
- Check freezer temperatures regularly.
- Place frozen food deliveries in freezers as soon as they have been inspected by an employee of the operation.
- Ensure good air flow inside freezers.
- Defrost freezer units regularly.
- Clearly label food prepared on site when it is intended for frozen storage.
- Keep the unit closed as much as possible.
- Schedule regular preventive maintenance.

Several dry food storage procedures are important:

- Keep storerooms cool and dry between 50°F and 70°F (10°C and 21°C).
- Ensure that all storerooms are well ventilated.
- Store dry food away from walls and at least 6 inches off the floor.
- Keep dry food items out of direct sunlight.

Storage procedures for specific food items are found in the National Restaurant Association's *ServSafe Coursebook.*

STORAGE LOCATIONS

Locations for product storage vary in operations with different production volumes. For example, very small operations typically have one storage location for refrigerated, frozen, and dry food items. There may be a work station refrigerator in which all refrigerated items are stored, with production staff working out of this refrigerator throughout the shift. There may be an upright or chest-type freezer, and a small room or even an open-access shelving unit for dry storage. This storage arrangement can work well when there are frequent product deliveries and when dependable vendors minimize the need for high inventory safety levels.

Larger-volume operations are more likely to have work station storage areas with additional backup storage. There may be a work station refrigerator for easy access during production with a several-door refrigerator located in another area. A larger freezer may also be available, and dry-storage areas will need to be larger than in smaller-volume operations, and preferably lockable.

OPEN FOR BUSINESS

KEEPING IT SAFE

Slips and falls are an ever-present safety challenge in all areas of the kitchen. They are of special concern near water sources such as pot and pan sinks and the dish-washing area.

Receiving locations create their own special problems. For example, floors may become wet as fresh poultry or seafood is removed from shipping containers filled with shaved ice. Sometimes ice is added in the receiving area after product check-in, or in storage areas after the containers are moved.

Those who receive products and who are responsible for them in storage must know how to implement safety precautions. Staff should mop the floors and place warning signs by wet spots as soon as possible. Dress codes that specify shoes with slip-resistant soles are another example of how to help ensure the safety of food handlers.

Still larger operations may have one or more walk-in refrigerators and freezers and a lockable dry-storage area. Efficiently designed facilities have some refrigerated and perhaps frozen-storage capacity in work station or production line work areas.

RECORD-KEEPING REQUIREMENTS

Inventory record-keeping systems are developed to help managers know the amount and costs of products in inventory. Without this information, they cannot calculate food costs. Accurate inventory systems are also needed to determine the quantity of and timing for products to be ordered. Recall that two systems for determining product purchase quantities, the minimum–maximum and par inventory systems, were discussed in chapter 3.

One additional way to manage inventory is to calculate the **inventory turnover rate**: the number of times each accounting period (usually a month) that the quantity of food or beverages in inventory must be purchased to generate the food or beverage revenue for the accounting period.

A formula is used to calculate the inventory turnover rate:

Cost of goods sold ÷ Average inventory = Inventory turnover rate

The **cost of goods sold (COGS)** is the cost to purchase the products that generate the food or beverage revenue.

This example shows how to calculate the inventory turnover rate for food. The inventory turnover rate for beverages is calculated in a similar way.

A manager conducts a physical inventory of the food products in storage areas on the last day of every month (using the procedures discussed in chapter 3) and learns the following from the July inventory count:

- Cost of food inventory at beginning of month = $39,500
- Cost of food inventory at end of month = $37,500
- Food cost (cost of goods sold) for the month = $88,000

With this information, the manager calculates the food inventory turnover rate using the following formula:

Cost of goods sold ÷ [(Beginning inventory + Ending inventory) ÷ 2]
= Inventory turnover rate

This can be performed in two steps. First, calculate the average monthly inventory:

($39,500	+	37,500)	÷	2	=	$38,500
Beginning inventory		Ending inventory				Average inventory

Then, calculate the inventory turnover rate:

$88,000	÷	$38,500	=	2.29
COGS		**Average inventory**		**Inventory turnover rate**

If the inventory turns over approximately 2.29 times per month, the food products in inventory will last about 13 days.

30 days per month ÷ 2.29 turns = 13.1

The ideal inventory turnover rate for any operation must be evaluated by considering the cost of inventory that managers determine should always be available. These decisions are based on concerns about the challenges incurred when there is too much or too little inventory. Some managers believe that a turnover rate below 1.00 times per month indicates excessive inventory levels and a rate above 3.00 may suggest an inadequate quantity on hand, risking stock outs.

While the inventory turnover rate for just one month may not be especially useful, the manager should note the changes in inventory turnover rates between fiscal periods. Why is the inventory turnover rate increasing or decreasing? What are the implications? What is the desired trend that the turnover rate should take? The answers to these and related questions can help operations better control their food inventory and the costs associated with it.

OPEN FOR BUSINESS

MANAGER'S MATH

1. What is the inventory turnover rate given the following information:

   Cost of food inventory at beginning of month: $64,750

   Cost of food inventory at end of month: $71,225

   Food cost (cost of goods sold) for month: $139,400

2. How many days will the food in inventory last (assume 30 days in a month)?

(Answers: 1. $\frac{\$139{,}400}{[\$64{,}750 + \$71{,}225] \div 2} = \frac{\$139{,}400}{\$67{,}988 \text{ (rounded)}} = 2.05$ times; 2. $\frac{30 \text{ days}}{2.05 \text{ times}} = 15$ days (rounded))

COSTS OF GOODS SOLD

Managers of every type of operation are concerned about meeting financial goals. Restaurant managers want to make a specified amount of profit. Managers in educational, business and industry, military, and other types of noncommercial foodservice operations (*Exhibit 4.7*) want to break even or to lose no more than a specified amount. Regardless of the type of operation, its financial success cannot be determined without knowing the amount of food cost, and this cannot be determined without knowledge of inventory costs.

Exhibit 4.7

Manager's Memo

All but the smallest-volume operations use an accrual accounting system that matches the revenue generated during a specific accounting period with the costs that were required to generate the revenue.

Food products may be paid for one month and used to generate revenue the next month. If the food cost is what was spent for food items during a month, it would be overstated for the month when the vendor's check was written because food was purchased but had not been sold. It would then be understated the next month because revenue was generated when it was sold but no cost for those products was incurred during the month.

For this reason, accrual accounting systems are used. Then the food cost reflects changes in inventory levels for purchased products, rather than just the cost of the products.

Managers calculate cost of goods sold (food or beverage costs) for two reasons:

- To compare the planned costs for food in their operating budget with the actual costs. An **operating budget** is a financial plan that estimates revenues and expenses for a specific time period. The budget's forecasted food revenues and food costs can be compared with actual food revenues and costs to help managers learn if any corrective actions are needed to bring future food costs more in line with costs estimated.
- To obtain the cost information required for the **income statement**, which summarizes the operation's profitability for the accounting period.

Exhibit 4.8 shows the calculations necessary to compute the cost of goods sold for food products at Sunset Shores Restaurant during January 2012. Note the use of inventory cost information on lines 1 and 4 in the calculations.

Exhibit 4.8

CALCULATION OF COST OF GOODS SOLD: FOOD (JANUARY 2012)

Line	Sunset Shores Restaurant		
(1)	Value of food inventory (beginning of period: January 1)	$83,575	
(2)	Value of food purchases (during January)	187,615	
(3)	Total value of food available (during January)		$271,190
(4)	Value of food inventory (end of period: January 31)		(89,540)
(5)	Unadjusted cost of goods sold: Food (January)		181,650
(6)	**Add adjustments to cost of goods sold: Food**		
(7)	Transfers from beverage		6,550
(8)	**Deduct adjustments to cost of goods sold: Food**		
(9)	Transfers to beverage	(4,175)	
(10)	Transfers to labor cost (Employee meals)	(8,900)	
(11)	Transfers to marketing	(3,750)	(16,825)
(12)	Cost of goods sold: Food (January)		$171,375

These calculations require the following information:

- Changes in inventory values (lines 1–5)
- An adjustment that increases cost of goods sold (line 7)
- Three adjustments that decrease cost of goods sold (lines 9–11)

Note that the value of food inventory at the beginning of the accounting period (line 1) is added to the value of purchases during January (line 2). This yields the total value of food available for sale during January (line 3). Then the value of inventory at the end of January (line 4) is deducted to determine the unadjusted cost of food used during January (line 5).

Some managers end their calculations at this point and consider the unadjusted cost of goods sold to be their food costs for the period. Other managers want to more closely match product revenue with product costs, so they make additional calculations.

In the bottom portion of *Exhibit 4.8*, the unadjusted cost of goods sold (line 5) is increased by the value of transfers from beverage (line 7). A **transfer** is an adjustment to cost of goods sold that increases or decreases food or beverage expense to match product costs with the revenue generated by the product's sale. Beverage transfers increase food costs because they generate food revenues. Examples of transfers from beverage include wine used in cooking and liqueurs used for tableside dessert flambéing. The beverage transfer represents costs initially charged to beverage (Cost of goods sold: beverage) which actually generated revenue for food (Cost of goods sold: food).

Exhibit 4.8 also shows some adjustments that reduce the cost of goods sold for food. Note the value of transfers to beverage (line 9), to labor cost (line 10), and to marketing (line 11). Transfers from food *to* beverage may include the cost of produce such as lemons and celery sticks used to make mixed drink garnishes, and ice cream used for specialty drinks. Transfers of food costs to labor cost represent the cost of employee meals, which is considered a labor cost. Transfers from food cost to marketing might include the cost of complimentary meals provided to potential guests looking for a site for a future banquet and to dissatisfied customers who receive a "comp" meal.

After all of these adjustments are made, the cost of goods sold for food (line 12) can be determined. The calculations to determine the cost of goods sold for beverages are very similar and involve transfers to and from food and to marketing. Note that, unlike food, there are no alcoholic beverages given to employees that would be treated as a labor expense.

SECURITY CONCERNS DURING STORAGE

The food and beverage products in storage were purchased with the intent that they will be used to generate revenue. If instead they are stolen, then money will have been spent and food or beverage expenses increased, but there will be no resulting revenue. Therefore, food or beverage costs will be greater than necessary, and profit levels will be lower.

Experienced managers consider their storage areas to be bank vaults, and the procedures they use for storage answer the question "How should money be managed in a bank vault?" Their answers to this basic question form the storage procedures they will use to address security concerns.

Basic procedures can help reduce employee theft. Examples include storing products in lockable areas with walls that extend to the ceiling, and limiting access to storage areas. With expensive products such as steaks and alcoholic

THINK ABOUT IT . . .

Some managers consider calculating inventory costs a "necessary nuisance." However, this information is used for budget control and income statements.

How important is taking inventory to most employees? What would you tell employees?

beverages, differences between quantities recorded in inventory records and quantities actually in storage should be investigated.

Some properties use employee package inspection programs to reduce the possibility of unauthorized employee "carry outs" (theft). A package inspection program is a policy that discourages employees from bringing backpacks, shopping bags, and other large packages to work and indicates that packages may be inspected when the employees leave work.

Some large-volume operations employ full-time storeroom clerks who, with only a few other employees, have access to storage areas. This staffing pattern is not practical for most operations, but managers of small-volume properties can use several strategies to limit employee access to storage areas.

For example, these areas can be kept locked with scheduled times during which the manager or other authorized person issues items from storage. Alternatively, expensive items can be kept locked and under perpetual inventory, and more general employee access can be permitted to the storage areas used for other items.

Wise managers know that problems can arise when employees have unlimited access to storage areas. As discussed earlier, they also understand that the problem is likely worsened when nonemployees such as vendor delivery or service representatives are allowed to enter.

PRODUCT ISSUING PROCEDURES

Product issuing procedures become important when food and beverages are moved out of storage and into production. The correct quantity of products must be issued to meet estimated production requirements. This process must be carefully controlled to minimize product misuse and enable managers to match issues of expensive items with the amount of revenues they should produce.

Importance of Effective Issuing

There should be some relationship between the quantity of food produced and sold, and the quantity removed from storage areas. Effective issuing practices help ensure that this happens.

Some managers allow any employee who needs something from storage to retrieve what is needed at any time. When this policy is used, every employee is really "in charge" of issuing. For example, production staff obtain food products, dining room servers may pick up containers of pre-prepared salad greens, and bus staff get the condiments needed to restock servers' areas. These examples fail to recognize the idea noted earlier of controlling storage areas as one would control a bank vault.

REAL MANAGER

CONTROLLING THEFT

In one of the restaurant operations, we were having a liquor cost problem. We were going through excessive amounts of Crown Royal and a very expensive brandy. I discovered, when I did my inventory, that we were going through twice as much as we had previously used. I compared the usage quantity to the sales quantity recorded on our point-of-sales system and found that they did not match.

We conducted inventory of the liquor in the morning, in the afternoon, and in the evening after closing. By doing so, we discovered that the liquor was disappearing after closing, but before the kitchen manager conducted the opening inventory. The only people who had access to the liquor during this time were the cleaning crew. When we discovered the shortage, we did a search of the cleaning crew's trash and equipment and found a bottle of Crown Royal hidden in the trash for them to pick up when they left. This is an example of how a proper use of systems helps you identify where in the flow of products your shortages occur.

Basic issuing procedures continue the control processes used when products are purchased, received, and stored. If these procedures are not implemented, food and beverage costs can increase with no offsetting revenues, and profitability will be lowered.

> **THINK ABOUT IT . . .**
>
> In many operations, storage areas are kept unlocked during the workday and locked at closing. Does this practice make sense to you? Why or why not?

Steps in Product Issuing

In operations using a perpetual inventory system, an issue requisition is typically used as authorization to remove items. An **issue requisition** is a document that authorizes an employee to remove products from storage areas. *Exhibit 4.9* shows a sample issue requisition used to obtain food products required for standard recipes.

Exhibit 4.9

ISSUE REQUISITION FORM

Date: 5/5/2012

Item	Issue Unit	No. of Issue Units	Cost per Issue Unit	Total Cost
Rib-eye steak	6-oz portion	25	$6.67	$166.75
				$166.75

Issue authorized by: J. D. Sill

Comments:

The rib-eye steaks in *Exhibit 4.9* are included in the operation's perpetual inventory records. Therefore, an issue requisition is needed to authorize removal of the products from the storage area.

Exhibit 4.9 indicates that the issue unit for rib-eye steak is a 6-ounce portion, 25 of which are to be removed from inventory. Since the cost per purchase unit is $6.67, the cost of the 25 steaks issued is $166.75.

25 portions × $6.67 per portion = $166.75

At the end of the shift or day, the signed issue requisition form is the authorization for the removal of the 25 steak servings, and the perpetual inventory form would be adjusted to reduce the quantity of rib-eye steaks available by 25.

> **THINK ABOUT IT . . .**
>
> A perpetual inventory is a lot of work. One alternative is to identify expensive and theft-prone items and control these items only.
>
> What are examples of items that managers might control with a perpetual inventory system?

Exhibit 4.10

What happens if only 22 steaks were sold during the serving period for which the steaks were issued? There would be 3 steaks remaining.

25 issued − 22 sold = 3 remaining

In some operations, these 3 steaks remain in the work station storage area to be used first during the next shift. In other operations, the 3 portions are returned to the central storage location, and the perpetual inventory balance is increased by 3 steaks. The managers in this type of operation believe that products are better controlled when they are locked in central storage areas than when they remain in production areas.

The use of issue requisitions to remove just the most expensive and theft-prone items would be considered practical in many operations. However, large-volume establishments with full-time storeroom staff may require that all items leaving storage be included on a signed issue requisition. In all operations, issue requisitions should be signed by a manager or other authorized employee (*Exhibit 4.10*), and the products and quantities issued should be based on reasonable estimates of production needs.

Wise kitchen managers realize that food products must be effectively managed at every step from purchasing through customer service. Issuing procedures are sometimes overlooked, but doing so can increase food and beverage costs.

TECHNOLOGY AND RECEIVING, STORING, AND ISSUING

Computer technology can assist with product receiving, storing, and issuing in several ways. First, "paperwork" can be eliminated, and this can reduce the communication problems that often occur when documents flow through several departments and persons within each department.

Traditional hard-copy purchase orders specifying the operation's purchasing needs and commitments can be electronically routed to receiving staff. Wireless technology allows receiving employees to check incoming products against the purchase order without needing to print a copy. Purchasers in large-volume operations can request bar code labels on incoming containers so quantity and cost information can be scanned and automatically entered into the inventory management system. Establishments of all sizes with bar code readers and software programs can use bar code technology to assist with purchasing and inventory management.

Technology enables electronic routing of purchase specifications, delivery schedules, and other communications between the food-production, purchasing, accounting, and receiving department staffs. Also, information from daily invoices can be electronically summarized on daily receiving reports, which generate information used for daily food costing calculations or other purposes.

Some operations use bar code systems to assist with product issuing. If bar codes have been attached to products in storage, items can be scanned when removed to generate a variety of information:

- Inventory quantities are adjusted.
- Purchasers are alerted if the inventory level has been reduced to the order point.
- A report can be issued about the estimated revenue to be produced by the items issued. For example, if 20 steaks are issued with a sales value of $15 each, $300 in revenue should be generated (20 steaks × $15 per steak = $300).

Managers can take end-of-month physical inventories very quickly and accurately when bar codes are attached to products in inventory. They must scan only the bar codes on each container to electronically determine the quantity and cost of all products in inventory. This information can be transferred to other systems to complete the operation's financial statements.

OPEN FOR BUSINESS

RESTAURANT TECHNOLOGY

Managers in many operations, especially small ones, are discovering that technology can assist with management, control, and accounting responsibilities.

A manager's first thought may be that only large-volume operations use this technology, and manual systems work adequately. It is interesting to note that, perhaps less than 20 years ago, many operations did not use the automated point-of-sale (POS) systems that are in use in operations of every type and size today. Benefits of these systems apply to operations of all sizes, and system pricing has been reduced to make them cost-effective for even small-volume operations.

It will be interesting 20 years from now to learn the extent to which automated systems for controlling products during receiving, storing, and issuing are in use in properties of all sizes. It is likely that machines will do more in restaurant and foodservice operations than they currently do. If this is true, managers can then focus their attention on areas where machines are not effective.

SUMMARY

1. **Explain each of the steps in an effective product receiving process.**

   Effective receiving requires staff who are properly trained to perform receiving tasks. They must know how to recognize proper quality and complete required delivery documents. The receiving area must be well planned and located, if possible, close to where deliveries will be made. Adequate space is needed to check incoming products. Tools and equipment such as pocket thermometers, a platform scale, and transport equipment are also needed.

   Steps in product receiving include comparing the delivery invoice and purchase order, confirming product quality including checking temperature, signing the delivery invoice, and moving products to storage. A final step involves completion of the receiving report.

   Receiving staff must use proper food safety practices and must carefully check all incoming products to confirm that the operation receives all of the products it is paying for. If products do not meet standards, a credit memo should be issued.

2. **State procedures for maintaining product quality, reducing the possibility of stock outs, and controlling costs during storage.**

   Dry, refrigerated, and frozen storage areas must be maintained at the proper temperatures. Cleanliness is important, and several food safety practices must be implemented to help maintain product quality during storage.

   Storage locations and capacities often relate to production volumes and are affected by the frequency of deliveries.

Effective managers routinely calculate inventory turnover rates and determine the costs of products in inventory. Then they can calculate the monthly cost of goods sold (food or beverage cost) for comparison with the operating budget and completion of the income statement.

Managers should control products during storage as they would control money in a bank vault. Storage areas must be physically secure, and a perpetual inventory system that tracks the quantities of expensive and theft-prone items is important.

3. **Describe procedures for controlling product costs when products are issued from storage to production areas.**

   Effective issuing helps ensure that products removed from storage generate the expected amount of revenue. Most operations do not have full-time storeroom staff. Instead, they develop practical issuing systems that closely control the most expensive items. Issue requisitions are used to access perpetual inventory products that are kept under lock. Employees may then have access to other products without an issue requisition. Nonemployees should never have access.

4. **Explain ways that technology can assist with receiving, storing, and issuing tasks.**

   Technology eliminates much of the paperwork involved in processing information related to receiving, storing, and issuing. Incoming products can be checked against electronic purchase orders without the need to print copies. Bar code systems allow product quantity and cost information to be automatically issued into and removed from the property's inventory management system. Electronic versions of purchase specifications, delivery schedules, and communication among employees involved with purchasing enable the process to flow smoothly. Technology may be increasingly helpful even to properties of relatively small production volumes.

APPLICATION EXERCISE

Break into teams of three or four students. Your team is an experienced restaurant or foodservice consulting group hired by the owner of a relatively small Italian restaurant. She employs three full-time cooks and several part-time assistants, and the owner also works in the kitchen during busy shifts.

The operation serves about 650 customers each week with a check average of about $12 per customer (food only).

The menu features seafood and meat main dishes, as well as two different steaks for regulars that consistently order these items. High-quality cheeses are an important ingredient in many recipes.

Her market is a highly competitive one. To be able to offer the best prices for the value received, her costs must be streamlined. The owner has turned to your group to identify the most efficient ways to control costs during the receiving, storage, and issuing processes.

Your assignment is to recommend practical, cost-effective, no-cost and low-cost product receiving, storing, and issuing systems.

Review the chapter to determine procedures the owner might implement at each of these control steps. Develop a report to explain your recommended system to the owner.

REVIEW YOUR LEARNING

Select the best answer for each question.

1. **Which document transfers product ownership from the vendor to the property?**
   A. Credit memo
   B. Purchase order
   C. Delivery invoice
   D. Purchase requisition

2. **When products are delivered, the quality of incoming products should be compared to the**
   A. purchase specification.
   B. purchase requisition.
   C. purchase order.
   D. credit memo.

3. ***Slack-out* is a term used to describe what type of seafood?**
   A. Seafood that is dried and canned
   B. Seafood that was frozen, then thawed
   C. Seafood that is alive when delivered
   D. Seafood that is spoiled

4. **What is a backorder?**
   A. An order that was left off the delivery vehicle
   B. An order that will be split into two deliveries
   C. An order that a vendor cannot fill immediately
   D. An order that is created because of buyer error

5. **What is the minimum recommended dry-storage temperature?**
   A. 40°F (4.5°C)
   B. 50°F (10°C)
   C. 60°F (16°C)
   D. 70°F (21°C)

6. **Refrigerated storage should be held below which temperature?**
   A. 39°F (4°C)
   B. 41°F (5°C)
   C. 47°F (8.3°C)
   D. 51°F (11°C)

7. **If the inventory turnover rate is 3.00 per month, approximately how many days will the inventory last?**
   A. 6
   B. 8
   C. 10
   D. 12

8. **What is a primary reason that managers calculate cost of goods sold?**
   A. To check it against purchase order information
   B. To determine whether the menu needs revision
   C. To compare it with operating budget information
   D. To consider whether purchase specifications are accurate

9. **Why do some managers make adjustments to the cost of goods sold?**
   A. To determine whether employee theft has affected costs
   B. To evaluate the effectiveness of make-or-buy decisions
   C. To better match product costs with revenue generated
   D. To determine whether inventory shortages exist

10. **Which document authorizes an employee to remove products from a storage area?**
    A. Broken case report
    B. Receiving report
    C. Issue requisition
    D. Transfer

5 Quality Food-Production Standards

INSIDE THIS CHAPTER

- Establishing Quality Standards
- Quality and Environmental Concerns
- Production Planning
- Production Methods to Enhance Quality
- Food Safety and Quality Standards
- Supervising Production Staff
- Other Production Quality Concerns

CHAPTER LEARNING OBJECTIVES

After completing this chapter, you should be able to:

- Describe basic concerns in establishing quality standards.
- Explain sustainability and "farm-to-fork" concerns.
- Describe how to determine menu item quantities and labor hours needed for production shifts, and indicate necessary instructions to production staff.
- Explain how food-production methods enhance quality.
- Identify food safety concerns that must be addressed as food quality standards are implemented.
- Identify supervision procedures during production that address quality and other concerns.
- Describe quality concerns when considering types of convenience food and when developing production procedures to enhance nutrition.

KEY TERMS

aging, p. 140
benefits, p. 133
classic sauce, p. 143
crew schedule, p. 133
cross-contamination, p. 146
farm-to-fork, p. 130
foodborne illness, p. 146
labor cost, p. 133
leavening, p. 145
line-up meeting, p. 135
make or buy analysis, p. 131
marinating, p. 140
master schedule, p. 134
mise en place, p. 135
production forecast, p. 131
production sheet (prep sheet), p. 136
roux, p. 143
sales forecast report, p. 135
sustainability, p. 129
TCS food, p. 138
tenderizing (meat), p. 140
thickener, p. 143

CASE STUDY

"I don't know why she looks over our shoulder so much," said Coline, a longtime cook at Brasserie Orchid. "The new manager has been here more than a month, and she should know that we always do our job—and we do it correctly."

"Yes, you're right," replied Molly, another cook. "I asked the manager why she does that, and she said it was a policy at her last property to make frequent checks of everything going on in the kitchen. She said that over the years, she had found many problems and corrected them before they became a big issue. Now she doesn't feel comfortable when she's not there in the kitchen managing—or whatever she calls it."

"I call it getting in the way!" said Coline. "I think she doesn't trust us, and she never will."

1. What kinds of things might the cooks be concerned about when the manager is in the kitchen?
2. What should the manager do when she discovers the cooks' concerns?

ESTABLISHING QUALITY STANDARDS

Customers in most areas of the country have numerous dining-out alternatives. If they want a steak dinner, they can pay a relatively small amount in some establishments and a lot more in others. The variation arises because restaurant and foodservice owners and managers establish quality standards, provide a meal experience meeting those standards, and then charge customers a price designed to generate a profit.

If Pat's Western Lodge offers the finest beef served by exceptionally well-trained employees in a luxurious setting, the price will be higher than at Family Cafeteria, serving customers with other preferences. Both of these operations can be successful. They are serving a different target market of customers who find value in different quality standards. Both properties will be successful if they offer a dining experience meeting their customers' standards at selling prices that provide value for the customer and a profit for the operation.

The concept of quality is hard to define. Some managers and customers simply perceive that a product or service is, or is not, of the proper quality. Unfortunately, an evaluation of food, service, and the dining environment cannot occur until a meal is completed.

While it is possible to assess whether quality standards *have been* met, the best managers plan ahead to help ensure quality standards *will be* met. This requires a commitment to quality at every step of the food management process. Managers think about quality at the same time that they think about standards. Quality is attained when standards have been met. The standards have been defined long before the customers arrive for a meal; managers know what their target market wants, and plans are in place to deliver it consistently.

Production planning is made easier when the standards, discussed in earlier chapters, are in place:

- The menu has been planned to meet target customers' food preferences.
- Standardized recipes are in place to provide food quality and portion sizes that represent value at the menu's selling prices.
- Food purchase specifications define the food quality requirements of ingredients purchased.

Properties that consistently deliver products and services at a value to customers attract a strong customer base. Operations that consistently *exceed* their customers' value expectations and provide even more than customers expect are highly successful.

THINK ABOUT IT . . .

Managers implement standards before production begins. Recall that purchasing, receiving, storing, issuing, and production are closely related. Each impacts the next.

What are examples of problems that can occur before products reach the kitchen that can affect production?

THINK ABOUT IT . . .

Some managers that exceed customers' expectations refer to this as the "wow factor." Flowers on Mother's Day, newspapers with breakfast, and a chef who visits tables are examples.

What additional "extras" might be wow factors?

QUALITY AND ENVIRONMENTAL CONCERNS

Today's restaurant and foodservice managers are increasingly concerned about **sustainability**: activities including water conservation, energy efficiency, and recycling that can lessen an operation's impact on the environment. They understand that successfully addressing these concerns is good for business as well as for the planet. The sustainability movement arose from an increasing awareness that buying locally produced food can be good for business, customers, and the local economy. Both of these concerns are important as production plans are made.

Sustainability

Some owners and managers believe their business can survive only if profit is their primary concern. It is true that an unprofitable business cannot survive. However, it is also true that "planet-friendly" practices yield many positive financial outcomes for businesses and for the economic health of their communities. There are many free and low-cost ways to practice sustainability during food production.

WATER CONSERVATION

Managers should take steps and train employees to conserve water while producing food:

- Reduce faucet use. Thaw frozen food products in the refrigerator instead of running water over them in the sink (*Exhibit 5.1*).
- Soak pots and pans and scrape dishes and cookware before washing them.
- Test and repair leaks. A faucet leaking just a tenth of a gallon a minute wastes 50,000 gallons or more annually.
- Keep water temperature at the right level. Use a thermometer to make sure water is not being heated more than needed. In some operations, employees must adjust the faucets to add cold water because the hot water is too warm.
- Install low-flow spray valves and save up to $1,000 a year. Local utility or water districts may offer rebates or repayment programs.
- Add aerators. Water-efficient aerators in kitchen sinks can save as much as a gallon of water per minute.
- Install on-demand water heaters. These cost-effective water heaters heat water only when needed.
- Replace steam cookers. Convection steamers use up to 90 percent less water than traditional models.

Exhibit 5.1

THINK ABOUT IT . . .

Typically, an operation spends 35 percent of its energy dollars on cooking, 28 percent on heating and cooling, 18 percent on dish washing, 13 percent on lighting, and 6 percent on refrigeration. What are some ways managers can reduce energy usage?

ENERGY CONSERVATION

There are many ways to reduce energy usage and costs and to increase the efficiency of kitchen equipment and facilities:

- Turn off lights in locations that are not in use.
- Ensure that ovens are full when they are used.
- Power down nonessential kitchen equipment when not in use.
- Reduce energy used for heating and cooling by sealing off unused areas and keeping exterior doors closed.
- Reduce idle time. Implement a set start-up and shut-down schedule for broilers, fryers, and ranges to reduce preheat and idle times.
- Develop an energy management and monitoring system.
- Improve your current maintenance program.

Farm-to-Fork Sustainability

Farm-to-fork describes the flow of food through the stages of growing, harvesting, storage, processing, packaging, and preparation. In other words, it is the path that food follows from those who grow or raise it to those who prepare and serve it. Ideally, this path will be environmentally friendly and short, so that freshness will be maximized and health risks minimized.

Customers are increasingly knowledgeable and concerned about the source of the food and beverages they buy. Consumer-related concerns about food safety, sustainability, freshness, and quality influence the farm-to-fork movement. Many customers want to support establishments that are operated in environmentally friendly ways.

Quality ingredients, when properly prepared, produce superior menu items. Managers recognize the advantages of locally grown food products:

- They are often fresher and may be of better quality than those processed and shipped from distant locations. This is especially so when the locally grown items are in season.
- They can often be purchased cost-effectively.
- They support family farming and the local economy.
- They are appreciated by many guests.
- They may reduce transportation costs, pollution, and excess packaging.

Interest in and concerns about locally grown food products should begin when menus are planned. Some managers feature numerous local items and use them as part of the establishment's brand. They may offer daily specials and use menu item descriptions that emphasize locally grown products when possible. These menu items and ingredients must then be considered as standardized recipes are developed and as products are purchased and produced.

Food-production staff must be trained to properly process and prepare these items to maximize food quality. **Make or buy analysis** involves procedures to determine whether menu items should be made with raw ingredients or purchased in a convenience form with some labor "built in." As alternatives are evaluated, managers should consider the purchase of locally grown food ingredients as an option.

PRODUCTION PLANNING

This section explains how production managers determine the quantities of products to produce and provides basic information about how the number of required labor hours is estimated. It also reviews the types of activities that must be done before production begins.

Determining Menu Item Quantities

Food-production managers develop **production forecasts** to determine the quantity of each menu item that will likely be sold during a specific time period, such as lunch or evening meal. This is an important management skill because an accurately developed production forecast suggests how many customers will be served and even what they will likely order. This information helps ensure that the right quantity of each menu item will be available.

Managers cannot predict exact sales for any future period, but they can estimate the number of each item they will sell. Managers can use information from the past, present, and future to create an accurate production forecast.

PAST SALES TRENDS

What has happened in the past is often a great indicator of what will occur in the future. If, for example, a manager knows the number of steak dinners sold on each of the last eight Fridays, he or she can begin to forecast steak dinner sales for this coming Friday.

Modern point-of-sale (POS) systems allow managers to keep excellent records of past menu item sales. A POS system is electronic equipment that records menu item sales and financial information on almost any basis that is helpful to the manager. Examples of historical menu item sales information helpful for production forecasting include the following data *for each menu item*:

- Yesterday's sales
- Average sales for a specified number of *same* days (e.g., Tuesdays)
- Last week's daily sales on a by-day basis
- Last two weeks' daily sales on a by-day basis
- Last month's daily sales on a by-day basis
- Actual sales on the same day of the last month or year

OPEN FOR BUSINESS

WHAT'S THE FOOTPRINT?

Restaurant and food-service managers can use a six-step recycling process:

1. Audit the trash to learn what is being thrown away. Identify what can be recycled.
2. Select someone to oversee the recycling program, track results, and regularly report progress.
3. Check with the waste hauler or recycler to learn if the operation can "single stream": whether all designated recyclable materials can be put in one bin for sorting at a recycling facility. If not, set up color-coded recycling bins or containers for paper, organics, and garbage. Post signs with pictures and simple instructions by recycle bins.
4. Co-op or consolidate. Recycling large amounts of glass, paper, and aluminum cans can generate substantial revenues from recyclers. Consider forming a co-op or recycling alliance with other businesses.
5. Create a recycling environment. Train employees, show them what to recycle, and involve customers.
6. Let customers and the community know how the operation is helping the environment.

Exhibit 5.2 shows a sample POS sales report that managers can generate to help estimate the sales of each menu item next Monday.

Exhibit 5.2

MENU ITEM SALES HISTORY FOR MONDAY, 2/20/2012

Menu Item	Last Monday Sales	Average Prior Five Monday Sales	Average Daily Sales Last Week	Average Daily Sales Last Month
Strip Steak	24	21	18	19
Salmon	8	9	10	9
White Fish	11	14	10	12
Grilled Chicken	35	30	32	29
Filet Mignon	45	47	42	49

CURRENT SALES TRENDS

Past sales information should be compared to recent sales data. Assume that a manager knows that for the last year, sales have increased about 5 percent each month from the same month the previous year. However, there has been no increase the last three months over previous years' sales. Does this mean the increase trend has slowed or stopped? The answer is that it appears the increase trend has slowed or stopped, but the manager should be very interested in reviewing sales over the next few months to confirm this. The best managers revise historical trends by considering current conditions.

Factors that may impact current sales trends include undesirable weather conditions or road construction near the establishment, which could reduce the sales forecast. Sporting events and festivals or other activities in the community could increase sales forecasts.

THINK ABOUT IT . . .

Future events can impact the number of customers, and this information is valuable for production planning. What types of events do you think influence restaurant and foodservice customer counts locally? Which will likely have the greatest impact?

FUTURE SALES TRENDS

Future conditions should also be considered when estimating menu item sales. Examples include the opening of new competitive establishments, planned promotions, and planned changes in hours of operation. Local newspapers, trade or business associations, and the chamber of commerce may have information about events and situations that could affect future sales levels.

After past, current, and future sales trends are assessed, managers can develop their production forecasts. An example is shown in *Exhibit 5.3*.

Exhibit 5.3

PRODUCTION FORECAST FOR MONDAY, 2/20/2012

Menu Item	Last Monday Sales	Average Prior Five Monday Sales	Average Daily Sales Last Week	Average Daily Sales Last Month	Production Forecast for Monday, February 20
Strip Steak	24	21	18	19	25
Salmon	8	9	10	9	10
White Fish	11	14	10	12	15
Grilled Chicken	35	30	32	29	37
Filet Mignon	45	47	42	49	50

Note in *Exhibit 5.3* that the manager reviewed available information and then estimated the number of each menu item to prepare for Monday, February 20.

Determining Labor Hours

Managers must schedule the right number of employees in the right positions at the right times to produce the menu items and services meeting quality standards expected by customers. They must do this while staying within labor cost goals. **Labor costs** are the money and cost of benefits paid for the work that employees do. **Benefits** are indirect financial compensation paid to attract and retain employees. Some benefits are required such as social security, and others are voluntary, like vacation days.

Managers must use an effective scheduling process to determine which employees will be needed at specific times to serve the number of expected customers. The **crew schedule** that results is a chart that shows employees what days and hours they are expected to work during a specific time period, usually a week.

Effective scheduling is always a challenge because schedule planners cannot always predict business flow. For example, there can be business volume peaks when a large number of customers arrive at the same time, and business volume valleys where few customers require food production and service. Managers must try to anticipate this and are challenged to have the correct number of employees available (not more or fewer than needed) to match customer needs.

Another scheduling concern relates to normally slower business towards the end of the dining period. On the one hand, fewer production and service personnel are needed to meet customer requirements, but employees are still needed for end-of-shift cleanup responsibilities.

Details of an effective employee scheduling process are explained in chapter 7. An overview is provided here to show the types of concerns food-production managers have as they plan for production.

Managers can use four steps to plan the number of labor hours needed for specific work shifts.

Step 1: Determine How Much Money Can Be Spent

Managers should not spend more for labor costs than the amount planned in the budget. Assume that the June labor budget for hourly paid production employees in an operation open only for dinner is $38,900 including benefits. This means the manager can spend an average of only $1,295 per day:

$38,900	**÷**	**30**	**=**	**$1,295 (rounded)**
June labor (wage) budget		**Days in June**		**Average daily wage**

If the average hourly wage is about $15, no more than 86 labor hours can be scheduled for an average day in June:

$1,295	**÷**	**15**	**=**	**86 (rounded)**
Average daily wages		**Average hourly wage**		**Average waged hours per day**

Also, no more than 602 waged hours can be worked in an average week:

86	**×**	**7**	**=**	**602**
Average waged hours per day		**Days open per week**		**Average waged hours per week**

Step 2: Create a Master Schedule

A **master schedule** considers the expected number of customers and allocates the number of hours allowed by the budget (step 1) among the different positions. It shows the number of employees needed in each position and the total number of hours each employee must work.

If the production manager is expecting 150 customers for a specific meal period, he or she may know from experience that four cooks working a total of 20 hours will be needed for that shift. The manager's experience will also help determine the number of persons in other positions and the hours they must work.

Step 3: Develop a Crew Schedule

A crew schedule tells employees the days and hours they are expected to work for a specific time period. Before it can be planned, employees must tell managers about the times during that period when they cannot or do not want to work. They do so by making time-off requests for vacations and other purposes.

Each employee's abilities should be considered when schedules for specific shifts are planned. More experienced and productive employees are often scheduled during times of high business volumes. However, it is not a good idea to schedule all of the best employees for one shift and leave the other shifts with only weaker crew members.

Legal concerns about scheduling minors must be followed, and care is needed to avoid scheduling overtime. Overtime is the number of hours of work per week, usually 40, after which employees must receive a higher pay rate. The overtime rate is usually 1.5 times the basic hourly rate.

Crew schedules should be fair. For example, employees who work at the beginning and end of shifts may have setup and cleanup duties. These duties are shared when all employees work these shifts on different days.

Step 4: Distribute and Adjust the Crew Schedule

Crew schedules should be distributed approximately 7 to 10 days before the first day of the schedule period. They can be provided in several ways, such as posting on employee bulletin boards (*Exhibit 5.4*), including with paychecks, emailing to employees, and posting on the establishment's intranet site.

Exhibit 5.4

Before-Shift Instructions

Food-production employees must work as a team, and this involves effective planning and communication before production begins. Many production managers begin work shifts with a **line-up meeting**. This meeting provides specific information for the shift including special events, new ingredients, or discussion about how to resolve an emerging or recurring problem.

The daily **sales forecast report**, a report that tells the production staff about the overall product needs to meet the sales forecast, may also be reviewed. The last column in *Exhibit 5.3* shows this information for a specific shift. It is especially important to share this information with staff if the volume of food production is unusual.

There is typically much *mise en place* necessary at the beginning of a work shift. ***Mise en place*** is a French term that means "get everything in place." Staff are much more productive when they assemble items they will need before they start preparation. For example, while preparing for the shift, production staff members might ask themselves what food items must be pre-prepped, and how they should be prepared. Should items be chopped, cut, sliced, diced, or julienned? How much product must be prepared? Their pre-shift activities will be driven by the answers to these questions.

Exhibit 5.5 shows a sample lunch preparation sheet. A **production sheet (prep sheet)** indicates items that will be needed for the shift.

Exhibit 5.5

SAMPLE LUNCH PREP SHEET WITH PAR LEVELS

Prep Sheet for Grill station

Item	On Hand	Par Level		Initials
		Mon–Thu	Fri–Sat	
8-oz steaks		50	80	
12-oz steaks		40	100	
New York strip steak		25	35	
Burger patties		90	150	
Vegetables		1 half-size hotel pan	2 half-size hotel pans	
Shrimp skewers		60	120	
Green pepper sauce		3 pints	3 quarts	
Au jus		1 quart	3 quarts	
Habanero mayonnaise		1 pint	2 pints	
BLT setups		90	150	
Cheese (Swiss, mozzarella, colby, jack)		30 slices each	90 slices each	
Duck glace		1 pint	3 pints	

The par level is an important part of the prep sheet. A par level system considers the quantity of products needed to bring the inventory level to an allowable maximum (par). It shows the quantity of ingredients needed to produce the number of menu items required for the shift. For example, notice that the prep sheet in *Exhibit 5.5* specifies that 50 8-ounce steaks and 1 quart of *au jus* are normally needed Monday through Thursday.

The par level changes depending on the day of the week, and when special events are scheduled. Note that different par levels are needed for weekdays and weekends.

Par levels are based on recipe yields. Many establishments produce slightly more sauces and dressings than needed to allow for portions that may become unusable and to ensure there are always enough on hand.

The prep sheet is used daily at the beginning of each shift. Typically, cooks remove food from storage areas and enter information on the prep sheet about the amount of product on hand. The production manager then verifies and signs off on the par level amounts.

Production managers may also create a food-production sheet for the shift. The sample in *Exhibit 5.6* indicates each menu item, recipe number, quantity needed, and special information.

Exhibit 5.6

PRODUCTION SHEET

Daily Production Sheet

Day	Date	Weather	Events	Customer Forecast
Monday	7/30/12	Mild, 75°F	Birthday party	200

Item	Recipe Number	%	Prepare	Sales	Leftover
Strip Steak	M-623	35%	70		
Salmon	E-328	11	22		
White Fish	G-929	25	50		
Grilled Chicken	G-930	18	36		
Filet Mignon	K-730	11	22		

Standardized recipes should be reviewed and distributed. In some operations, recipes are maintained at individual work stations and are always available. At other properties, computerized recipes are generated to yield the specific number of servings required for the shift.

It is common practice for specific employees to prepare the same menu items each shift based on their abilities and job descriptions. Even if this is done, employees should be cross-trained so they can produce other items if they need to. **Cross-training** increases an employees' ability to perform tasks other than those they regularly do.

After these beginning-of-shift activities are completed, food production can begin.

PRODUCTION METHODS TO ENHANCE QUALITY

Entire books have been written about food-production activities. This section highlights some basic procedures.

Exhibit 5.7

Fresh Fruit and Vegetables

Fresh vegetables must be thoroughly washed as a first step in preparation (*Exhibit 5.7*). Potatoes to be used for baking should be scrubbed with a stiff vegetable brush. Leafy green vegetables should be washed in several changes of water. With each change, the greens should be lifted from the water so soil and other additives can sink to the bottom. After washing, vegetables should be drained well in a colander, covered, and refrigerated.

Pasta and potato salads have been involved in foodborne illness outbreaks, so care is needed when preparing them:

- Make sure leftover TCS food used to make salads has been handled carefully. **TCS food** is food that needs time and temperature control for safety.
- Prepare food in small batches so large amounts of food do not sit out at room temperature for long periods of time.
- Consider chilling all ingredients and utensils before using them to make the salad.
- Leave food in the refrigerator until all ingredients are ready to be mixed.

Fresh produce must be handled carefully to prevent foodborne illnesses:

- Make sure fruit and vegetables do not come in contact with surfaces exposed to raw meat and poultry.
- Refrigerate and hold cut melons, cut potatoes, cut tomatoes, and cut leafy greens at 41°F (5°C) or lower.

Vegetables can be served raw, such as celery and carrots, or they can be cooked using several of the same methods used for meat, fish, and poultry.

Consistently cut vegetables look appealing, and cooks can more accurately calculate cooking times. *Exhibit 5.8* shows classic vegetable cuts.

Exhibit 5.8

CLASSIC UNIFORM VEGETABLE CUTS

	Basic Shape	Name	Size
	Round	Rondelle disk-shaped slices	Varies, but each piece should be the same size. Cut perpendicular to the food.
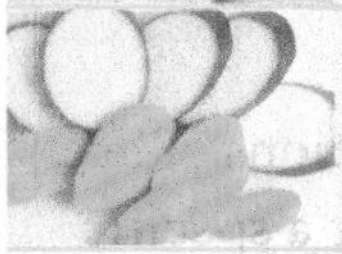		Diagonal: oval-shaped slices	Varies, but each piece should be the same size. Cut at an angle to the food.
	Stick	Batonnet	2″ x ¼″ x ¼″
		Julienne	2″ x ⅛″ x ⅛″
	Round/ Square	Paysanne	Basic shape depends on type of vegetable. Pieces are uniformly ½″ x ½″ x ⅛″
	Cube	Large dice	¾″ x ¾″ x ¾″
		Medium dice	½″ x ½″ x ½″
		Small dice	¼″ x ¼″ x ¼″ (start with Batonnet)
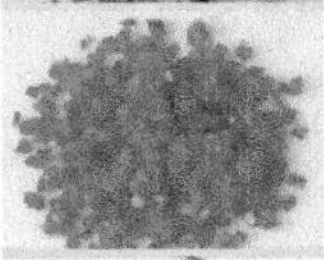		Brunoise	⅛″ x ⅛″ x ⅛″ (start with Julienne)
	Thin Strip	Chiffonade	Roll leafy greens or herbs into a tube, then cut into thin strips.

Meat, Fish, and Poultry

Meat preparation processes enhance flavor or make tougher grades of meat more tender.

Tenderizing refers to breaking down connective tissue in meat. These methods do not require liquid:

- **Scoring:** Small cuts are made across the surface of thin slices of meat.
- **Pounding:** Special hammers are used to pound cuts of meat.
- **Blade or needle tenderizing:** Meat is passed through a machine with multiple blades or needles that penetrate it.

Marinating is another tenderizing method. It typically involves immersing food items from one to several hours in liquid made from oil and vinegar or lemon juice seasoned with herbs and spices. Marinating is typically used with a dry heat cooking method such as grilling or broiling.

Aging can tenderize beef. Wet aging involves placing small portions of beef in a plastic bag without air, sealing the bag, and refrigerating for a specific period of time. Dry aging is done in a special cooler, usually by specialty vendors using large cuts of beef. Air is circulated at precise temperatures for the time determined by package size and other factors.

There are several other meat processing methods:

- **Curing:** The most common curing method uses salt, a combination of salt by-products, sugar, herbs, spices, and flavored alcohol.
- **Smoking:** Smoking adds the flavor and smell of the material used to create the smoke.
- **Brining:** Meat is soaked in salted water, usually with sugar and sometimes spices. This is typically done in a plastic bag that is sealed and refrigerated for 1 hour per pound.

The objective of any cooking method is to make a product tender. It also affects the product's texture and flavor and can reduce the number of microorganisms that are present. Products cooked too long will become tough or lose nutrients, appearance, and quality. There are three basic cooking methods for meat, seafood, and poultry. The product, the temperature to be applied, and the cooking time affect selection of a cooking method.

METHOD 1: MOIST HEAT

Food is cooked in liquid (water or stock) or just above the liquid:

- **Boiling:** Food is submerged in liquid maintained at the boiling point, 212°F (100°C).

- **Simmering:** Liquid is maintained at a hot but not boiling temperature and usually bubbles slightly.
- **Poaching:** Food is completely covered with liquid (submerged poaching) or placed on vegetables and half-covered with liquid (shallow poaching).
- **Steaming:** Food is cooked above a hot liquid without the liquid touching the food.
- **Blanching:** Food is cooked briefly in hot water and cooled very quickly (the food may not be thoroughly cooked).

METHOD 2: DRY HEAT

Heat is applied without any water, and fat may be used:

- **Broiling:** High heat is applied from above, usually to tender cuts of meat and some vegetables.
- **Grilling:** Similar to broiling, but heat is applied from below.
- **Baking and roasting:** Food is cooked in an open dish or on a rack in an oven. Baking usually refers to breads and pastries, while roasting refers to meat.
- **Sautéing:** Meat, usually thinly cut, is quickly cooked with a small amount of fat in a very hot pan on the stovetop.
- **Frying or pan-frying:** Similar to sautéing, but with a moderate amount of fat.
- **Deep-frying:** Food is submerged in hot oil for a short amount of time, until it turns brown. Deep-frying does not tenderize most meat and is not often used on beef, veal, lamb, or pork.

METHOD 3: COMBINATION METHOD

Combination cooking is often used to make tougher cuts of meat tender, and to give the meat better flavor:

- **Braising:** A whole roast of meat is seared (browned) on all sides at high heat. After searing, the meat is covered about halfway in a liquid, sometimes red wine, and slowly cooked with root vegetables.
- **Stewing:** Similar to braising, but the meat is first cut into small pieces and then browned. Excess fat is removed and a cooking liquid such as wine, stock, or water is added. The pot is kept covered, and the liquid should not boil.

Exhibit 5.9 shows the different cooking methods along with types of food suited to each method. It also indicates food items that can lose quality if cooked with each method.

Exhibit 5.9

ADVANTAGES/DISADVANTAGES OF COOKING METHODS FOR DIFFERENT FOOD

Cooking Method	Good for	Bad for
Moist Heat	*Overall, results in a more delicate product, except for the blanching method.*	
Boil	Dense, tough products such as shoulder cuts of meat, and hardy root vegetables	Already tender products, such as fish and asparagus, or products with a high water content
Simmer	Fish and delicate vegetables or fruit	Dense, tough meat products
Poach	Fish, poultry, and delicate vegetables	Dense, tough meat products
Steam	Fish and vegetables, especially greens	Dense, tough meat products
Blanch	Vegetables, especially greens	Dense, tough meat products
Dry Heat	*Overall, results in more tender protein and more sturdy produce.*	
Bake/roast	Products with a high water content, whole fish, and seafood	Tough cuts of meat or leafy green vegetables
Broil	Meat products, fish, sturdy vegetables, and certain fruit	Soft or too-small products
Deep-fry/fry	Products with high starch content and low water content	Tough cuts of meat or leafy green vegetables
Grill	Meat products, fish, sturdy vegetables, and certain fruit	Soft or too-small products
Sauté	Universal, but for some products it may be necessary to finish the cooking process with a moist heat method	
Combination	*Overall, enhances very tough protein and high-fiber produce.*	
Braise	Tough cuts of meat and hardy root vegetables; can be used for large pieces of meat	Already tender products or products with high water content
Stew	Tough cuts of meat and hardy root vegetables; meat is usually cut into small pieces	Already tender products or products with high water content

Sauces

When well chosen and prepared, a sauce brings out the best qualities of the food it accompanies. Sauces are liquid-based and are generally made with a thickening agent.

THICKENING AGENTS

A **thickener** is any ingredient added to a liquid to make it thicker. One type of thickener is a **roux** made from equal parts flour and butter, cooked on the stove at a low temperature. A roux is used in many sauces.

Slurry is another thickening agent. It is a combination of cold liquid and cornstarch mixed and then brought to a simmer. Some sauces are finished by a liaison: a thickening agent made of heavy cream and egg yolk. Modern sauces (those that are not classic) are thickened with arrowroot, a starch obtained from a tropical herb. It is the only thickening agent that can be frozen and defrosted without losing texture

CLASSIC SAUCES

The **classic sauces** are called "mother sauces," since other sauces can be made from them. There are five classic sauces, and each takes much practice to make well:

- *Béchamel sauce* is made from flour, butter, and milk.
- *Velouté sauce* is made from a chicken, seafood, or veal stock thickened with a roux. It is used as a base for cream soups.
- *Espagnole (brown) sauce* is made from meat stock plus brown roux (roux made with fat) and mirepoix: a combination of cooked vegetables, usually one-half onion, one-quarter celery, and one-quarter carrot.
- *Hollandaise sauce* consists of egg yolks beaten into a liquid such as lemon juice or vinegar. Whole butter or clarified warm butter is gradually beaten in until the consistency is a thick liquid.
- *Tomato sauce* uses fresh or canned tomatoes and may be thickened by simmering over low heat for a few hours.

SMALL AND MODERN SAUCES

Small sauces are made using any classic sauce as the base and then adding different flavorings. Examples include various cream sauces, horseradish, and châteaubriand.

Modern sauces do not use butter and eggs and are becoming more popular, especially with nutrition conscious diners. Examples include chutney (cooked relish), salsa (chopped vegetable or fruit sauce), and coulis (puréed fruit or vegetables).

Dairy Products

Cow's milk and products made from it are included on most menus. Basic information about dairy products is very helpful as production procedures are considered.

Exhibit 5.10

BUTTER AND CHEESE

Butter (*Exhibit 5.10*) is made by separating the fat in fluid milk and chilling the resulting product. It must include at least 80 percent milk fat, and salt and coloring may be added. Butter smokes and burns at fairly low frying temperatures, but its flavor and popularity for cooking and for table-service uses make it a popular in many operations.

Cheese is a concentrated dairy food made from milk, and numerous varieties are available. Processed cheeses are a blend of cheeses that have been shredded, mixed, heated, and then molded. Their ingredients may include wine, fruit, vegetables, and meat.

MILK AND CREAM

Beverage milk is sold in several forms and must be pasteurized before sale. Pasteurization involves heating milk products to a temperature high enough to destroy potentially harmful bacteria. This reduces food safety problems and extends the time that the product can be safely stored. Cream is the high-fat liquid part of whole milk.

These are the most common forms of milk and cream:

- **Whole milk:** Must contain a minimum of 3.25 percent milk fat.
- **Lower fat milks:** Made by reducing the fat content of milk. Common forms include 2 percent milk fat, 1 percent milk fat, and skim milk, which has less than 0.05 percent milk fat.
- **Nonfat dry milk:** Made by removing almost all fat and water from pasteurized milk. It contains about half the calories of whole milk.
- **Half and half:** Must contain at least 10.5 percent but not more than 18 percent milk fat.
- **Light cream (coffee or table cream):** Must have at least 18 percent but less than 30 percent milk fat.
- **Light whipping cream:** Must have at least 30 percent but less than 36 percent milk fat.
- **Heavy (whipping) cream:** Must have at least 36 percent milk fat.
- **Sour cream:** Produced by adding a special bacteria culture to light cream. It is smooth and thick and meets milk fat requirements for light cream.

FROZEN DAIRY PRODUCTS

Frozen dairy products include ice cream that is made from milk, cream, sweeteners, and flavorings, with at least 10 percent milk fat. Frozen custard is made by adding egg yolks to ice cream. Low-fat ice cream may contain no more than 3 grams of fat per 4-ounce serving and is made by adding egg yolks to regular ice cream.

Food preparation using dairy products ranges from simply serving the item (ice cream, milk, or coffee creamer) to the complex process of whipping various creams to create delicate soufflés. Using the freshest products that have been properly stored and handled and carefully following standardized recipes will yield quality products.

Bakery Products

Bakery products are made from batters and doughs. They include muffins, biscuits, pastries, cakes, cookies, and breads. They are an important part of many menus, but few operations produce all bakery items on-site.

Managers typically make purchase decisions based on factors including quality, price, portion size, availability, and perishability. Since bakery items are highly perishable, the amount to be purchased is important.

Baked products use grain flours made from wheat (most often used), corn, rice, or other grains. During milling, wheat kernels are ground, and their parts (bran, germ, and endosperm) are separated and recombined to make the four common types of wheat flour:

- **Bread flour:** Made from hard wheat with a relatively high protein content and used to produce yeast breads and rolls.
- **All-purpose flour:** Used in many baking applications including yeast breads, quick breads, and cakes.
- **Pastry flour:** Used for pastries and cookies.
- **Cake flour:** Used to make fine-textured cakes and other delicacies.

Some customers are allergic to wheat and gluten, and production and service employees should be aware of this. Gluten is protein in wheat that gives structure to baked products containing wheat flour.

In addition to flour, food-production staff work with two other common ingredients to prepare baked products:

- **Sugar:** A sweetening agent that enhances flavor and softens the texture of baked goods by breaking down gluten: a protein found in flours. Common sugars are granulated (table sugar), confectioners' (powdered sugar), brown (sweetened with molasses), and chocolate (made from dry cocoa powder).
- **Leavening agents:** Leavening is the process in which ingredients produce gases that cause dough to rise. Common leavening agents are yeast (fresh or dry), baking soda, and baking powder.

Manager's Memo

It is important to maintain proper food storage temperatures. Refrigerated food storage should be 41°F (5°C) or below.

Set freezers to a temperature that will keep products frozen. This temperature will vary from one product to another.

Pathogens (disease-causing micro-organisms) grow best at temperatures of 41°F to 135°F (5°C to 57°C). The time food products are exposed to these temperatures including receiving, storing, preparation, production, and holding should be less than four hours. An important caution: This includes the time before products reach the operation's back door. That is why it is important to purchase from professional suppliers.

FOOD SAFETY AND QUALITY STANDARDS

For an operation to stay in business and be successful, guests and employees must feel safe. Therefore, no management responsibility is more important than the need to protect the health and well-being of customers and employees.

Importance of Food Safety

Each year, thousands of cases of foodborne illness are reported, and it is likely that many thousands more are not reported. **Foodborne illness** is a disease transmitted to people by food. It can be caused by pathogenic microorganisms, chemicals, or physical hazards.

Managers know that their operation's reputation will be affected if a foodborne illness occurs. Information may be broadcast by local news media, and local health departments increasingly post inspection results on their Web sites. Lawsuits may be filed by alleged victims, and legal costs and lost time can result.

Food Safety before and during Production

Frozen food products should not be thawed at room temperature or left in a sink full of water. There are several acceptable ways to thaw frozen food:

- In the refrigerator at a temperature of 41°F (5°C) or lower
- Under running potable (drinkable) water at a temperature of 70°F (21°C) or lower
- In a microwave oven if food will be cooked immediately afterward
- As part of the cooking process if the product reaches the required minimum internal cooking temperature

Food items in production should be kept at room temperature for the shortest time possible. Do not remove all items from storage at the beginning of a shift.

Utensils, such as knifes and cutting boards, and work counters should be properly cleaned between food-preparation tasks to prevent **cross-contamination**: the transfer of microorganisms from one surface or food to another. Frequent and proper hand washing is necessary during food preparation and all other foodhandling. Also, accurate thermometers must be used to monitor food temperatures.

Food Safety after Production

When cooling hot food, food should first be reduced in volume by placing it in smaller containers. Then one or more of several effective methods can be used to cool hot food quickly and safely:

- Placing food in an ice-water bath
- Stirring food with an ice paddle
- Placing food in a blast chiller or tumble chiller
- Adding ice or cold water as an ingredient

Some menu items such as steaks cut to an approximate uniform weight may be served immediately after production, and their time in the temperature danger zone is slight. However, items such as casserole dishes and sauces may be held before service, and this must be done at a temperature above 135°F (57°C).

Some operations provide food items on self-service salad bar, dessert bar, or buffet counters. Food held in public areas must be kept at 135°F (57°C) or higher or at 41°F (5°C) or below. These self-service areas must be kept clean, and sneeze guards or food shields (barriers to protect food on self-service counters from customers who might sneeze or cough) are required.

Cleaning and Sanitizing Practices

Additional steps to reduce foodborne illnesses include cleanup activities. Cleaning involves removing food and other waste matter from a surface such as a countertop. Sanitizing involves reducing the number of microorganisms on a surface to safe levels.

As shown in *Exhibit 5.11*, there are five steps for cleaning and sanitizing food-production and serving surfaces:

- Scrape or wipe the surface.
- Wash the surface.
- Rinse the surface.
- Sanitize the surface.
- Allow the surface to air-dry.

Equipment used for food production and service must be cleaned by following the manufacturer's instructions. Basic strategies include using a clean cloth, brush, or scouring pad and warm soapy water. Clean from top to bottom or from one side to another, and then rinse with fresh water. An approved sanitizing solution can be spread or sprayed onto food-contact services, and areas should be air-dried before use.

Cleaning chemicals can be dangerous and must be handled carefully. Hand-washing sinks must be conveniently located and kept supplied with liquid or powdered soap. Bar soap should not be used.

Exhibit 5.11

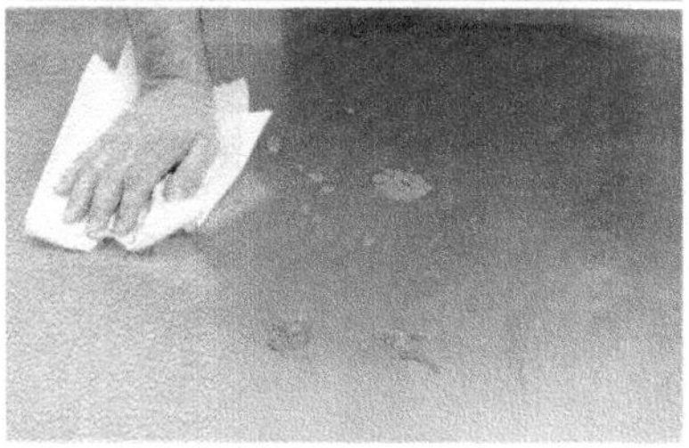

Scrape or wipe

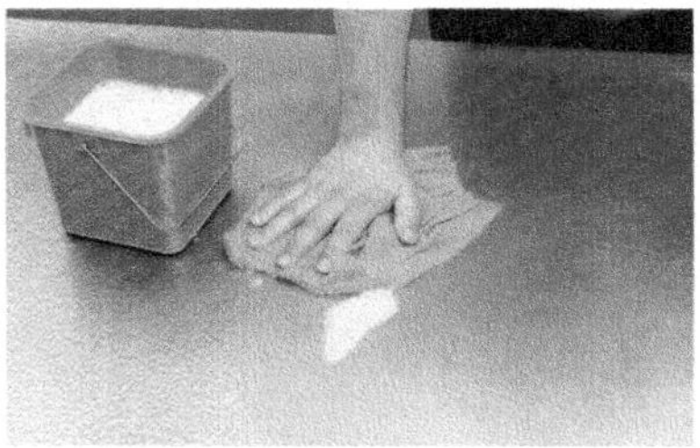

Wash

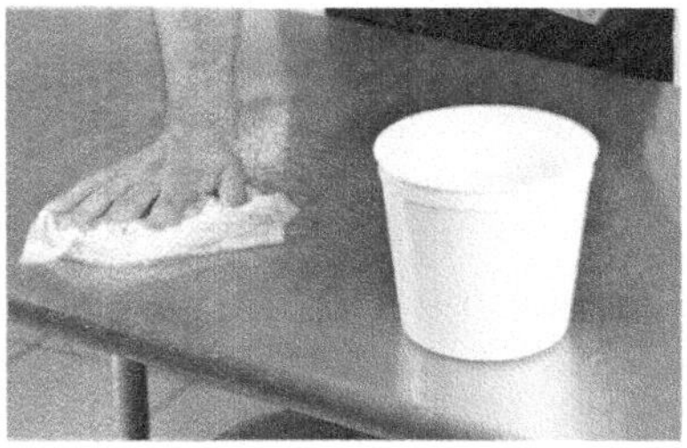

Rinse

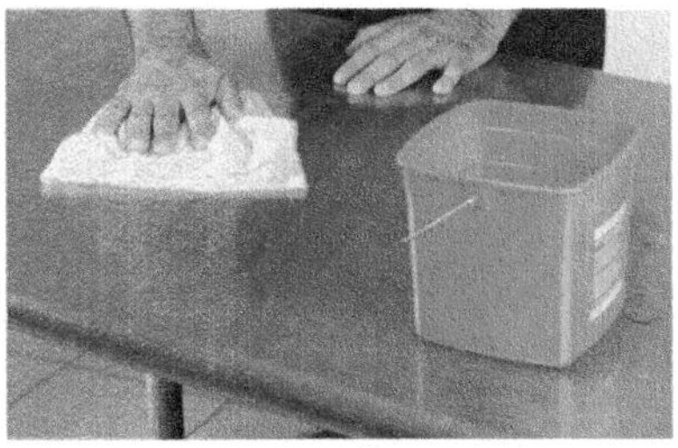

Sanitize

Air-dry

Other Food Safety Issues

Production managers are concerned about food safety concerns relating to chemical, physical, and pest control hazards.

CHEMICAL HAZARDS

Harmful chemicals such as toxins (poisons) in some shellfish occur naturally. Others that are human-made, including preservatives in food, cause illnesses in some persons.

Pesticides are chemicals used to kill insects, and they are applied to many fruit and vegetables before harvest. These chemicals may be on products when they are received, so proper washing is important.

Other chemicals, such as those used for facility and equipment cleaning, can get into or onto food and can cause illness and even death. Storing cleaning items away from food products and carefully labeling and using them are important handling procedures. Toxic metal poisoning can also occur if food items are stored or processed in some containers, such as those made from galvanized metals.

PHYSICAL HAZARDS

Food can be contaminated with physical hazards. Examples include fragments from glassware broken in food-preparation areas, metal shavings from can openers, and wood splinters from toothpicks or skewers used in food production. Other examples include human hair and pieces of labels from food containers. Foodhandlers must carefully inspect food items being produced and must avoid wearing unnecessary jewelry that can be lost in food during production.

PEST CONTROL HAZARDS

Pests, including rats, flies, and cockroaches, carry microorganisms and can contaminate food. It is important to prevent their entry into the building and also to eliminate food, water, and places to hide in case they get in. Effective pest control programs ensure that preventive actions are working.

Food should be properly stored, and garbage should be kept covered until removal, which should occur frequently. Containers should be cleaned regularly. Tight-fitting doors, screens, or air curtains should be used to reduce entry of flying insects.

Cracks, small holes, and other areas where pests may enter must be repaired. Food and supplies entering the building should be checked when received for signs of infestation. Signs of rodent infestation include droppings, burrows along walls and under garbage, gnawing marks in wood, and tracks in dust.

Unless someone in the operation has received special training, chemicals to control pests should be applied by professionals to reduce the chance that residues remain in or on food storage and preparation areas.

SUPERVISING PRODUCTION STAFF

Restaurant and foodservice operations must ensure that employees are following best practices. Managers must monitor for food and beverage quality and other concerns as they supervise production employees.

Supervising for Quality

Food and beverage production managers have a great deal to do as they supervise employees. This is an important responsibility, because planning to ensure that standards will be met will do no good unless plans are implemented consistently.

The manager must "manage by walking around" and observe preparation activities to ensure that standardized recipes are being followed and that all food safety procedures are in place. Knowledge of the standardized recipes and details of the property's HACCP plan are important. The manager should routinely monitor product temperatures to ensure that they are correct, using the correct thermometer for the food type.

The thermometer may be the single most important tool available to protect food. There are several types:

- **Bimetallic stemmed thermometer:** Measures temperature (typically from 0°F to 220°F [−18°C to 104°C]) through a metal probe with a sensor toward one end.
- **Thermocouple or thermistor:** Measures temperature through a metal probe or sensing area and displays results on a digital readout. These thermometers may be fitted with different types of probes.
    - **Immersion probe:** Designed to measure temperature of liquids, such as soups and sauces.
    - **Surface probe:** Measures temperature of flat cooking equipment like griddles.

Managers should observe the cooking line during both slow and busy times and compare work procedures to standardized recipe requirements, or disposed of if single use utensils are used.

Production managers often sample products and recipes to ensure that the food meets quality standards. To do so, place the food to be tasted in a separate dish and do the tasting with a clean utensil. The dish and utensil should then be removed from the food-preparation area for cleaning and sanitizing (unless they are disposable, in which case they should be thrown away or recycled).

Managers want to know if customers are enjoying their food. Dish-washing areas can be observed to determine whether excessive food is being left on plates, or dish washers can be asked. Food being returned to the kitchen is another indicator, as are ongoing discussions with the dining-room manager and service staff. If problems are noted, they should be corrected quickly, and managers should determine if the applicable standardized recipes are being correctly and consistently used.

Use of take-home containers is another concern. If this is being done frequently, could portion sizes be too large? Could smaller portions be available at lower prices to provide better value for customers? The review of invoices for purchase of take-home containers can be useful for noting trends in take-home consumption.

If food or beverage quality problems are observed, they must be resolved quickly. First, confirm that the correct standardized recipe is being followed closely. Temperature logs from the HACCP system should be checked. Problems can be caused by equipment, lack of communication, untrained employees, and other factors.

When solutions to food quality problems are implemented, it is important to monitor revised procedures closely to ensure they are effective. Revised procedures should be incorporated into training programs to help avoid future food quality problems. In addition to training, tool and equipment changes may be necessary. Managers should ask "What do staff need to know or do to avoid continued food and beverage quality problems?" Their answer to this question provides direction for resolving the problems.

Unfortunately, sometimes quality-related problems cannot be resolved quickly. Problems are sometimes traced to human resources concerns such as selection, training, team building, and motivation. Often, signs of these problems extend beyond food quality to other areas of the operation. When this occurs, other actions are required that typically involve supervision or human resources activities.

Supervising for Other Concerns

The design of some kitchens makes the supervision of food-production employees difficult. Examples include large spaces, full walls, and multiple levels. However, managers must monitor employee output. Are standardized recipes being followed? Are employees preparing the items assigned on the menu production sheet (see *Exhibit 5.6*)? This is not normally a problem in a work culture where employees are encouraged to ask for assistance when needed. However, new employees and those with less experience require monitoring until the manager believes less direct supervision is needed.

Serving involves transferring food items from production to service employees. The restaurant or foodservice manager and food-production and dining-room managers and their teams should establish procedures for ordering that ensure standards are met:

- Menu items are ordered properly with no confusion.
- Menu orders are placed at the appropriate time.
- Complete orders are prepared at the right time.
- Production and service employees communicate and interact professionally.
- Bottlenecks that slow the serving process are identified and corrected.
- Customer complaints are addressed while the customer is still at the establishment.

Problems that are identified must be corrected. Some problems may occur only during a specific shift. Examples include producing and serving meals when the operation is short-staffed, and being unable to produce menu items because of broken equipment or unavailable ingredients. Experienced managers can deal with these problems, at least to some extent, and the best managers step in and assume some line responsibilities when they need to.

Longer-term problems include food always coming out slowly when several large groups order at once or too many customers order specific items at about the same time. These problems might be addressed in the first case by reservation and seating standards and in the second instance by careful menu review. In both cases, problems will be resolved by food-production and service teams working together to determine procedures that benefit customers.

At the end of the shift, the manager must decide what, if any, unserved items can be reused. Proper food leftover procedures must be followed as hot food is safely cooled. The number of servings that can be reused will affect servings to be prepared the next day. Records of leftovers must be maintained and used to develop the next day's prep sheets and production records.

It may also be necessary to review inventory levels of some ingredients to ensure that the number of servings needed for the shift can be prepared. This is not normally a problem when an effective inventory management system is used (see chapter 3). However, if there are insufficient ingredients for the shift's production, the manager may be able to order more by telephone or email or leave a message for the next shift's manager.

THINK ABOUT IT . . .

Managers walk a fine line to have items available for all customers. If too few items are produced, they sell out. If too many, there are leftovers. How would you deal with these concerns?

Exhibit 5.12

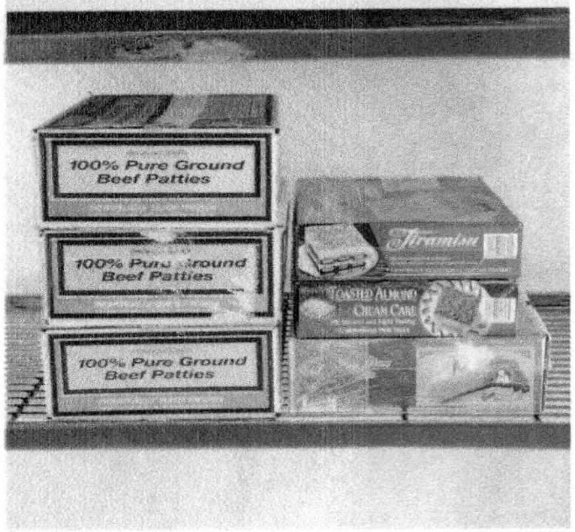

OTHER PRODUCTION QUALITY CONCERNS

Food-production managers understand that types of convenience food must meet quality standards. Managers must also understand that customers increasingly desire food produced in a way that addresses nutritional concerns.

Types of Convenience Food

Convenience food items are those in which labor has been added to reduce the amount of labor in the manager's own kitchen (*Exhibit 5.12*). Examples include preformed hamburger patties to eliminate the need for portioning, and loaves of sliced bread to eliminate on-site baking. Restaurant and foodservice managers must consider whether it is in the best interest of their business to buy a convenience item or to produce the item on-site.

Not all convenience items are of good quality. Just as managers should avoid purchasing raw ingredients that result in poor-quality items, they should also avoid using poor-quality convenience items. It is important to learn about the convenience items that may be useful to the operation, and they should not be purchased unless they meet the property's quality requirements.

THINK ABOUT IT . . .

Managers who use types of convenience food can offer a varied menu without extensive production. Consider the purchase of several quality fruit pies as opposed to making pies on-site.

What are advantages and disadvantages of convenience food products?

Nutrition and Food Production

Managers incorporate nutrition concerns into menu planning, purchasing, receiving, and storing activities. However, nutrition goals cannot be attained unless they are addressed during the food-production process.

Food production begins with the development of standardized recipes. Ingredients in every recipe can be evaluated to determine their nutritional impact:

- Butter used to brown meat during cooking might be eliminated if a nonstick pan is used.
- Margarine might be used in a recipe specifying butter.
- The amount of cheese used in a casserole item might be reduced without sacrificing product quality.
- Tofu and yogurt might be used to thicken some food products.

Standardized recipes should be carefully followed as ingredients are weighed and measured. Also, while portion control requirements imposed by standardized recipes should always be followed, they are a special concern for items being served to nutrition-conscious customers.

Possibilities also exist for increasing the nutritional quality of many food items as they are produced.

MEAT, POULTRY, AND FISH

Consider the following suggestions for improving the nutritional content of meat, poultry, and fish:

- Trim visible fat before preparation.
- Use lean (lower-fat) ground beef and other beef products.
- Emphasize cooking methods such as roasting, broiling, or grilling rather than frying.
- Use fat-free or reduced-fat products.
- Remove the skin from poultry before roasting or frying.
- Strain or skim fat from soups, stews, and other products that contain meat. Fat can be skimmed from the top of products after they are chilled for storage when the fat congeals and rises to the top.
- Use marinades without excessive fat, salt, or sugar. Herbs, spices, lemon juice, and yogurt can often provide good flavor and higher nutritional qualities.
- Roast at low temperatures to reduce product weight and nutrient loss.
- When stir-frying, ensure that the fat is hot before adding the food; less fat will be absorbed.
- Use unsaturated fats in a spray form to reduce the amount of fat used when cooking.
- Broil or grill the food suspended so fat can drain off.

FRUIT AND VEGETABLES

Use these fruit and vegetable preparation suggestions:

- Recognize that peeling, cutting, and soaking reduces water-soluble vitamins in items processed.
- Clean and process vegetables close to cooking and service times.
- Cook for the minimum time necessary.
- Use as little fat or oil as necessary when sautéing.
- Use as little water as possible when cooking.
- Use the water remaining after fruit and vegetables are boiled.
- Steam vegetables whenever possible. This will reduce nutrient loss by almost 50 percent compared with boiling.
- Bake root vegetables in their skins rather than after peeling, cutting, or boiling.

Fruit and vegetable salads are increasingly popular because more people understand the nutritional value of the primary ingredients. In addition, note that fat-based dressings such as Thousand Island and Russian are not always necessary for delicious salads. Vegetable dressings, nut-based dressings, and olive oils can enhance taste and be used lightly. Cream and egg yolk–based dressings can be diluted with broth or skim milk, and puréed fruit and vegetable juices can also be ingredients in salad dressings.

OTHER NUTRITIONAL PREPARATION SUGGESTIONS

Some customers are concerned about nutrition when ordering entrées and reward themselves by ordering desserts. Creative production employees can make desserts nutritious:

- Offer smaller portions.
- Use only egg whites or reduce the number of egg yolks.
- Use vegetable shortening instead of butter or lard when baking.
- Replace fat with applesauce (in some baking recipes).
- Use skim, low-fat, and nonfat dairy products instead of heavy cream or whole milk.
- Use low-fat spreads, jellies, and jams in dessert recipes.
- Flavor with vanilla or other extracts and spices instead of sugar.
- Offer fruit in different forms, such as baked or poached, as desserts.

SUMMARY

1. **Describe basic concerns in establishing quality standards.**

   Managers must establish standards that define the experience that customers will receive. Then they must charge a value-driven price that generates a profit for the operation. The best managers use planning tools including the menu, standardized recipes, and purchase specifications to help ensure standards will be met.

2. **Explain sustainability and "farm-to-fork" concerns.**

   Managers understand that addressing sustainability will be good for their business as well as the planet. They train their employees to conserve water and energy. They also consider farm-to-fork concerns that minimize the flow of food from growing to preparation. Locally grown food has advantages, and increasingly, customers prefer it.

3. **Describe how to determine menu item quantities and labor hours needed for production shifts, and indicate necessary instructions to production staff.**

   Production forecasts based on past, present, and future sales trends suggest the quantity of each menu item to be sold. Required labor hours are determined by considering the number allowed by the operating budget and scheduling based on the number of customers. These calculations allow master and crew schedules to be developed.

Production managers should conduct line-up meetings with employees who must assemble items for preparation. Much of this information is included in a prep sheet. A production schedule indicating responsibilities may also be distributed.

4. **Explain how food-production methods enhance quality.**

   Basic preparation procedures should be developed for handling fresh fruit and vegetables; meat, fish, and poultry; sauces; and bakery and dairy products. Sanitation and other requirements should be addressed in standardized recipes with special attention given to TCS food items: those that need time and temperature control for safety.

   Cooking methods that apply moist heat, dry heat, or a combination of both can be used depending on the product, temperature, and cooking time.

   There are five classic sauces, as well as additional small and modern sauces, that can enhance food. Production employees must know quality considerations for butter, cheese, milk, and cream and for ingredients typically used in bakery products.

5. **Identify food safety concerns that must be addressed as food quality standards are implemented.**

   There are numerous procedures that should be used during and after food production to reduce the possibility of foodborne illnesses. Facility cleanup practices are also important, as are concerns for chemical, physical, and pest control hazards.

6. **Identify supervision procedures during production that address quality and other concerns.**

   Managers must "manage by walking around" to note activities related to quality. These include use of standardized recipes, compliance with the HACCP program, and work procedures. Managers know how to tell if customers are enjoying their food. They also establish procedures for transferring food items from production to service staff, and they can identify and correct short- and longer-term problems.

7. **Describe quality concerns when considering types of convenience food and when developing production procedures to enhance nutrition.**

   Convenience food items reduce the amount of labor needed to produce menu items on-site, but they must meet quality requirements.

   Numerous food-preparation procedures can address customers' increasing concerns about nutrition, and many can be incorporated into standardized recipes. Examples include substituting nutritional ingredients for less nutritional items and using appropriate cooking methods.

APPLICATION EXERCISES

Exercise 1: Line-Up Meetings

Break into teams of three or four students to plan and deliver a line-up meeting role-play.

First, each student should spend several minutes developing a list of topics that a food-production manager might discuss in a line-up meeting. Consider the line-up meeting topics mentioned in the chapter, but be creative and think of some others. Also, plan a 1- to 2-minute training session that will be part of the meeting. You might discuss a new equipment item being purchased, a new supplier who will be providing some products, or a new menu item that will be produced for the self-serve salad bar or buffet line.

Each student should then make his or her line-up presentation to the other students in the role-play team.

Exercise 2: Healthy Food Production

Do an Internet search for "nutritional food preparation" or "healthy cooking." Make a list of 10 suggestions for healthy food production that were not mentioned in the chapter.

REVIEW YOUR LEARNING

Select the best answer for each question.

1. **When should an establishment's quality standards be defined?**
   A. When the customers' meals are completed
   B. At the time the customers' meals are served
   C. Before the customers arrive at the establishment
   D. When employees decide how to do their jobs

2. **What is an advantage of aerators in kitchen sinks?**
   A. They reduce the possibility of foodborne illness.
   B. They produce the preferred water for thawing food.
   C. They save approximately a gallon of water per minute.
   D. They use approximately 20% of the energy dollar.

3. **What is an advantage of locally grown food products?**
   A. They often have uneven quality.
   B. They reduce transportation costs.
   C. They contain no preservatives or pesticides.
   D. They have almost 50% less nutrient loss.

4. **How can a POS system help with production planning?**
   A. It calculates par levels to meet sales forecasts.
   B. It provides data that can be used to develop food-production forecasts.
   C. It specifies the number of food-production employees needed for each shift.
   D. It indicates the amount of money that will be available to purchase food products.

5. **To develop a master schedule, the manager needs to know**
   A. employees' average hourly wage.
   B. employees' average overtime wage.
   C. availability of specific employees.
   D. abilities of specific employees.

6. **Food can be tenderized by**
   A. frying.
   B. baking.
   C. marinating.
   D. grilling.

7. **Blanching is which kind of cooking method?**
   A. Dry heat
   B. Moist heat
   C. Open-fire heat
   D. Combination method
8. **Which is a classic sauce?**
   A. Hollandaise
   B. Chutney
   C. Salsa
   D. Arrowroot
9. **What percentage of milk fat is required for a product to be classified as butter?**
   A. 50%
   B. 60%
   C. 70%
   D. 80%
10. **Which is a safe way to thaw frozen food items?**
   A. On a prep table at room temperature
   B. In a refrigerator at a temperature of 41°F (5°C)
   C. In a microwave oven if the food will be cold held
   D. In a sink full of water at a temperature of 70°F (21°C)

FIELD PROJECT

1. Make an appointment to visit a restaurant or foodservice manager or chef to discuss food safety concerns during food production. Ask the following questions and take notes of the responses.
   A. What are several of the most important procedures that you tell your foodhandlers they should always follow during food production?

   ____

   ____

   B. Based on your experience, what are the three most common food safety violations during food production?

   ____

   ____

   C. What food safety related factors are of most concern as you think about your property's work environment, work stations, and foodhandling equipment?

   ____

   ____

6 Quality Beverage Management Standards

INSIDE THIS CHAPTER

CHAPTER LEARNING OBJECTIVES

After completing this chapter, you should be able to:

- Explain the importance of providing quality non-alcoholic beverages.
- Describe federal, state, and local regulations that impact the purchase and management of alcoholic beverages.
- Explain basic management practices applicable to selecting, purchasing, receiving, storing, and issuing alcoholic beverages.
- Explain beverage-production standards for bar layout and drink preparation procedures.
- Describe the use of manual and automated beverage-production methods.
- Describe three beverage service methods.
- Explain procedures for managing wine sales.

KEY TERMS

bag-in-box (soft drink syrup container), p. 163

behind-bar par level, p. 172

bin number (wine), p. 183

call brand (spirits), p. 169

cellar temperature, p. 184

CO_2, p. 162

control state (alcoholic beverages), p. 168

decaffeinated coffee, p. 161

free pour, p. 176

herbal tea, p. 162

jigger, p. 178

license state, p. 168

liquor license, p. 165

measured pour spout (alcoholic beverage bottle), p. 178

off-premise license, p. 165

on-premise license, p. 165

overpour, p. 176

pasteurization (beer), p. 168

public bar, p. 180

sommelier, p. 184

underpour, p. 176

vintage, p. 183

well brand (spirits), p. 169

wine and food pairing, p. 182

wine list, p. 183

CASE STUDY

"Almost everybody who comes in and orders a drink wants Scotch!" said Denise, a bartender at Greener Meadow Lounge. She was talking to Carlos, the bar manager.

"We could really increase our business and make our customers happy if we increased the number of Scotches we offered," she continued. "In fact, I have an idea: Let's pick a busy night and ask every person who orders Scotch what his or her favorite brand is. This would be a random sample, and then we could order every brand suggested and start advertising our Scotches. We could even have a sales promotion called 'Dare You to Order a Scotch We Don't Have!' What do you think?"

1. If you were Carlos, how would you respond to Denise?
2. Carlos does want to increase the variety of Scotches. How should he determine what Scotches should be available?

INTRODUCTION

Much of this book has focused on the management of food products. Most food and beverage operations generate the largest percentage of revenues from the sale of food rather than beverage products. Also, the cost of food products is much larger than beverage costs in the typical establishment.

However, beverage sales are also important to many establishments. There are advantages to offering alcoholic beverages including popularity, profit, and the target market's expectations of enjoying wine with meals. But there are also several concerns such as possible liability if responsible alcoholic beverage service is not provided, and a loss of family appeal. In addition, numerous regulations, increased training and supervision costs for employees, and possible customer behavior problems can create challenges. Managers of operations selling alcoholic beverages must know how to effectively manage alcoholic beverage sales, and that is the primary topic of this chapter.

NON-ALCOHOLIC BEVERAGES

Before discussing alcoholic beverages, it is appropriate to review basic information about popular beverages that do not contain alcohol. While they are served by all restaurant and foodservice operations and are profitable, some managers take them for granted. This is unfortunate because the beverage's serving quality is impacted by how they are handled by the establishment's employees.

Exhibit 6.1

The number and type of beverages offered can be extensive, but this chapter will discuss those that are most common:

- Water
- Coffee
- Tea
- Soft drinks
- Milk

Water

Water is the world's most important and popular beverage. It is a uniquely American custom to offer complimentary water immediately upon seating customers. Managers can offer "tap" water available through the local water supply source, or bottled water. Some operations offer both, and many establishments offer both carbonated and "still," or noncarbonated bottled water (*Exhibit 6.1*).

This is of interest to managers who can now charge guests for a desired product that, at least traditionally in the United States, had previously been complimentary. Establishments that offer customers a choice of coffee, tea, or soft drinks may also want to sell one or more brands of domestic or imported bottled water.

The proper service of water is simple. However, it is still important to ensure that clean glassware is used, and that the water is properly iced or chilled. Glasses should be large enough to eliminate the need for a server returning to the table an excessive number of times. Additionally, if bottled water is served, the labels should be clean and clearly legible.

Coffee

Coffee is grown in many parts of the world and is enjoyed by restaurant and foodservice guests worldwide. Every manager should have a thorough understanding of the procedures needed to make a good cup of coffee. The process starts with the selection of a quality coffee bean to be ground at the operation, or the purchase of vacuum packed, pre-ground coffee from a reputable coffee supplier. The number of choices of brands, styles, and price levels is large. Managers must become familiar with the production process and service of both regular and **decaffeinated coffee**, which has had its naturally occurring caffeine reduced or eliminated entirely.

Quality coffee can be produced and served when attention is paid to the details:

- **Water quality:** Make coffee only with water that has been filtered just prior to use. Automatic coffee makers have filters built in; however, when using pour-through coffee makers, ensure that the water used comes from a filtered source.
- **Coffee bean quality:** The cost of coffee, like most other food items, increases as the product quality increases. Managers should buy the best-quality coffee they can afford to sell and then preserve it in storage until it is ready to be brewed.
- **Water temperature:** Each coffee maker is designed to brew coffee with a designated water temperature. Monitor the water temperature regularly to ensure the water used is neither too hot nor too cold.
- **Equipment:** Coffee-making equipment (*Exhibit 6.2*) must be kept very clean. Dirty equipment yields coffee with an off-flavor.
- **Holding time:** Coffee deteriorates when it is held for excessive periods of time. Many managers dispose of unused, brewed coffee after it has been held 30 minutes.

Exhibit 6.2

Tea

Tea is of three basic types: black, green, and oolong. In the United States, over 90 percent of the tea consumed is black tea. This has been fully oxidized, or fermented, yields a hearty flavor, and has an amber color. Some popular black teas include English breakfast, which is a good breakfast choice since its hearty flavor mixes well with milk, and Darjeeling, a blend of Himalayan teas suited for drinking with most meals. Orange pekoe is another black tea and is a blend of Ceylon teas. It is the most widely used of the tea blends.

Green tea is not oxidized, has a more delicate taste, and is light-green or gold in color. Oolong tea, popular in China, is partly oxidized and is a cross between black and green tea in color and taste. While flavored teas evolve from these three basic teas, **herbal tea** contains no true tea leaves. Instead, steeping the flowers, berries, peels, seeds, leaves, or roots of plants in boiling water creates herbal teas.

Making quality tea in a restaurant or foodservice operation is more difficult than making good coffee because the equipment to do so is not generally automated. Several concerns must be addressed to make good tea:

- **Water quality:** As is the case with coffee, use clean, fresh water with no off odors or taste.
- **Cup temperature:** Preheat the pot or cup that will be used for service. Fill the teapot or cup with very hot water and let it stand for a minute or two.
- **Water temperature:** Bring cold water to a boil in a microwave or use hot water from a coffee maker. The water should be at the boiling point (212°F or 100°C) when the tea or tea bags are added. If customers are served a pot of water and tea bag to make their own tea, the temperature of the water should be reduced to limit the chances of scalding if there is a mishap by the server or customer.
- **Steep time:** One to three minutes is best depending on the type of tea and water temperature.
- **Condiments:** A full array of condiments offered to the customer who orders tea includes granulated sugar, low-calorie or no-calorie sweeteners, lemon wedges, and milk.

Soft Drinks

Nearly all establishments serve soft drinks (*Exhibit 6.3*). These carbonated beverages are popular worldwide and generally are made by blending CO_2, purified water, and flavored sugar syrups. **CO_2** is carbon dioxide—the colorless, odorless gas used to carbonate water in soft drinks and beer. Many managers buy bag-in-box containers of syrup that are then mixed with water and CO_2 when the soft drink is dispensed. These 5-gallon syrup containers are so named because the soft drink syrup is sealed in a plastic bag that is then placed in a cardboard box for easy transporting.

Exhibit 6.3

Selecting the soft drinks to be sold in an operation usually is dependent on the vendor that provides or services the dispensing equipment used by the property. Operations that do not use dispensing equipment can select any brand of soft drink and sell it in single-serving bottles or cans.

Soft drinks produced and served on-site will be of high quality if attention is given to several factors:

- **Water quality:** Soft drinks are essentially flavored water so, as with coffee and tea, water quality is critical. Filtered water should be used.
- **Syrup levels:** Soft drink production equipment must be maintained and adjusted on a regular basis to ensure that the proper amount of syrup is mixed with the proper amount of water.
- **Carbonation levels:** Too much carbonation makes soft drinks taste bitter, and too little makes them taste "flat" and overly sweet.
- **Equipment cleanliness:** Soft drink dispenser heads must be inspected and cleaned regularly.
- **Ice levels:** Proper attention must be paid to the ratio of ice to soft drink so that customers will receive a properly chilled beverage without too much ice in the glass that displaces the soft drink.

Milk

Most operations that serve significant numbers of children have beverage milk on their menu. In addition, many adults prefer milk with their meals, especially at breakfast. Milk is also a common cooking ingredient. Today's health-conscious consumer may request something other than traditional "whole" milk, and managers may offer 2 percent, 1 percent, and skim milk alternatives.

The proper service of beverage milk is a simple, two-step process. The first step is to ensure the freshness of the product, and monitoring "use by" dates is critical. Second, serve the milk in clean glassware at the proper temperature of 38°F to 42°F (3°C to 5°C).

REGULATIONS AND ALCOHOLIC BEVERAGES

Managers serving alcohol should be aware that the alcoholic beverage industry is one of the country's most highly regulated. Federal, state, and local regulations directly affect many aspects of alcohol sales. Any restaurant or foodservice operation that sells alcoholic beverages must comply with a wide range of laws that govern how they are sold.

Federal Regulations

Most operations must comply with a variety of federal regulations such as the Americans with Disabilities Act (ADA), federal wage and hour laws, and regulations issued by the Equal Employment Opportunity Commission (EEOC).

Establishments that serve alcoholic beverages are also regulated by the Alcohol and Tobacco Tax and Trade Bureau (TTB). Its mission, in part, is to ensure that only alcoholic beverage products that have been properly taxed are sold. This agency, a division of the federal government's Department of the Treasury, also ensures that these products are manufactured and sold in accordance with applicable labeling, advertising, and marketing laws.

Managers interact with the TTB because their businesses must pay a special federal tax each year. After payment is received, a tax stamp verifying the taxes were paid is issued, and this stamp must be kept available for inspection at all times. Managers must also make the sales records, invoices, and receipts for alcoholic beverages purchased available if requested for TTB inspection.

The federal government's Bureau of Alcohol, Tobacco, Firearms and Explosives (ATF) also enforces laws applicable to alcoholic beverages. It is part of the Department of Justice and works in cooperation with other federal, state, local, and international law enforcement agencies. The ATF plays a variety of critical alcohol sales-related roles:

- It disrupts and eliminates criminal and terrorist organizations by identifying, investigating, and arresting offenders who traffic in contraband or illegal liquor, or both.
- It works to seize and deny further access to assets and funds used by criminal enterprises and terrorist organizations.
- It prevents encroachment of the legitimate alcohol and tobacco industries by organizations trafficking in counterfeit or contraband cigarettes and illegal liquor.
- It assists local, state, and other federal law enforcement and tax agencies to thoroughly investigate the interstate trafficking of contraband cigarettes and liquor.

Manager's Memo

Taverns have long been popular meeting places throughout the United States. By the end of the 1800s, some had evolved into the types of beverage operations in existence today including establishments, private clubs, and other entertainment operations.

The growth in alcoholic beverage consumption brought with it the temperance movement, which believed that drinking alcohol is morally wrong and harmful to the individual and society. In 1919, Congress passed an amendment to the Constitution that prohibited the manufacture, sale, transportation, and importing of alcoholic beverages. This law remained in effect until 1933, when Congress repealed it.

Individual states, counties, towns, and precincts now control the sale and use of alcoholic beverages by issuing spirits licenses. As a result, a wide variety of alcohol-related laws exist throughout the country. There is still a societal concern about the consumption of alcohol, and its use and sale and the establishments that serve it are highly regulated.

State Regulations

Businesses in all states must follow state laws that involve employee wage and hour and other human resources issues, building codes, business taxes, and a wide range of other topics. There are also state laws that apply to those selling alcoholic beverages.

A government agency in all states, typically called the Alcoholic Beverage Control (ABC) Commission, licenses operations to sell alcoholic beverages and monitors compliance with applicable laws. The ABC generally works in cooperation with county or city agencies, which may have their own regulations.

Several types of liquor licenses are available in most states. A **liquor license** is a state-authorized permit that allows the license holder to sell alcoholic beverages in compliance with state, local, and federal laws.

Establishments such as bars and restaurants typically have **on-premise licenses**, which allow the property to sell alcohol in the same location where it will be consumed. By contrast, grocery stores and other retail businesses have **off-premise licenses**, which allow them to sell alcohol that will be consumed somewhere else.

Establishment owners may request and be granted a license to sell beer only; beer and wine; or beer, wine, and spirits. The number of liquor licenses granted in a specific area is typically limited by the area's population. Liquor licenses are generally granted for one year and then must be renewed.

Licenses can be revoked if state regulations are violated. Managers operating businesses with a liquor license must be familiar with and carefully follow all regulations that apply to their operation. They must know about their responsibilities:

- **What is sold:** The type of license indicates what alcoholic beverages can be sold. In addition, the products can be sold only if they have been purchased from a state-approved alcoholic beverage supplier. This ensures product wholesomeness, allows the state to carefully monitor alcoholic beverage sales, and assists in the collection of applicable taxes.
- **Where it is sold:** Liquor licenses generally specify where the alcohol can be sold and consumed. This may be an individual building address or just part of a building. Managers should know the exact boundaries of the permitted area. In almost all cases, on-premise licenses will not allow customers to remove alcohol from the specified location. Some licenses may allow alcoholic beverage service in more than one location. For example, a license may permit beverage service at off-site catered events. In this case the license will clearly identify the situations in which off-site alcoholic beverage service is allowed.

- **When it is sold:** Operating hours for establishments that serve alcohol are strictly controlled. States typically regulate when alcohol may be sold, at what time service must stop, and the days of the week, holidays, or special occasions such as election days that service must be restricted or stopped.
- **How it is sold:** The responsible service of alcohol requires that managers implement procedures to control sales. For example, establishments may not be allowed to serve a customer more than one alcoholic drink at a time, or perhaps only those seated at a table or bar may be served.
- **To whom it is sold:** In all states and the District of Columbia, those who purchase alcoholic beverages must be at least 21 years old. Managers and their teams must take reasonable steps to ensure that they serve alcohol only to those legally entitled to make these purchases. However, managers are likely to encounter customers who are over 21 but are still prohibited from purchasing alcohol: those who appear to be intoxicated. Customers who are obviously and visibly intoxicated cannot be served alcoholic beverages. Serving such a customer can create significant legal and financial problems.
- **In what quantity it is sold:** Beer is typically sold by the bottle, glass, or pitcher. Wines can be sold by the glass, carafe, or bottle, and spirits are typically sold by the drink. However, local regulations may place restrictions on the quantity of alcohol that can be sold at one time to a specific individual.

In addition to ABC requirements, states are concerned about collecting the taxes due from alcoholic beverage sales. Records of alcohol purchases and sales must be maintained, and these records can be examined by the state's taxing authority.

Local Regulations

Many municipalities have their own liquor authorities and regulations that affect the sale of alcoholic beverages, and local laws are often stricter than the applicable state laws. These local authorities may be responsible for enforcing alcohol laws, issuing and monitoring compliance, issuing violations for citations, and holding hearings for violators of the liquor code.

Food and fire safety codes are also often developed and monitored by local governments and will be of concern to managers. Local health departments inspect all establishments selling food and beverages. Inspectors want to know how the establishment stores alcoholic beverages, washes its glassware, and maintains and stores juices, fruit garnishes, ice cream, and other items used to prepare alcoholic beverages.

Some highly publicized nightclub fires that caused many deaths in the early 2000s have made many community regulators increasingly concerned about fire safety in businesses serving alcohol. Local building codes related to construction materials, available building exits, and maximum occupancies may be enforced by police, fire, and building department personnel, and the zoning officials who regulate business locations.

THINK ABOUT IT . . .

Zoning requirements can impact signs, parking, and outdoor seating, especially for operations selling alcoholic beverages. Should requirements for operations selling alcoholic beverages be different from establishments that sell only food? Why or why not?

BEVERAGE MANAGEMENT PRACTICES

Many customers enjoy alcoholic beverages such as beer, wine, and spirits as part of their dining experience. A manager's concern about quality must extend to beverage aspects of the operation because procedures used for beverage production and service affect customers' views of the establishment.

Beverage operations can be profitable. For example, the contribution margin (product revenue minus product cost) for each dollar of beverage sales is typically much higher for beverages than for food.

Many procedures required to purchase, receive, store, and issue alcoholic beverages are the same as those used to manage food. Some, however, are different. Alcoholic beverages are expensive, theft-prone, and under significant regulations. Managers must implement practices to control alcoholic beverages and to ensure that what is actually spent for beverages is in line with planned expenses.

Selecting and Purchasing Beverages

Those responsible for purchasing alcoholic beverages must design a purchase system that considers several factors:

- The beverage products and bar supplies needed
- The quantities of products and supplies needed
- The quality requirements for necessary products
- The suppliers who will consistently supply the products at the best prices and on the most favorable terms
- The frequency of placing orders
- The persons who will be responsible for managing beverages

The answers to these questions establish the basic beverage purchasing system. The task is, in some ways, less difficult than purchasing food products. Wines, spirits, and canned and bottled beer are not highly perishable. Also, many products are purchased by brand, in which case the purchase specification is easily defined.

However, government regulations in many states affect sources of alcoholic beverage supply. In **control states**, the state is the sole supplier of spirits. Individuals and retail establishments must purchase all spirits directly from state stores. In **license states**, the state grants licenses to wholesalers, distributors, and sometimes to manufacturers that permit these businesses to sell alcohol within the state. The specifics of how much and what type of control is in place differs by state. For example, some control states retain their control of spirits sales but allow the licensing of beer and wine sales.

Spirits purchasing can be more complicated in license states than in control states because individual wholesalers do not carry all brands in all quantities. Distributors usually have exclusive sales rights for certain brands in specific areas. If a buyer wants a specific brand of scotch, for example, he or she can buy it only from a single distributor. The result is that no single supplier carries a complete selection of all available brands and items, and purchasers must order from several supply sources. Also, strict governmental control often results in minimum wholesale prices with limited discounting.

Regardless of the restrictions on purchasing procedures, managers must carefully select the beer and spirits they will offer. Wines are discussed later in this chapter.

BEERS

Many establishments offer a selection of traditional and lower-calorie (light) beers. The popularity of alternative beers varies by region, and local beer distributors will know the best-selling brands in the area.

Beer can be purchased in bottles, cans, or kegs. In nearly all cases canned or bottled beers are finished by **pasteurization**. This is a final step in beer production that heats the product to 140°F to 150°F (60°C to 66°C) for 20 to 60 minutes to kill any bacteria and remaining live yeast cells.

Some canned or bottled beer is not pasteurized. In this case, the beer is passed through very fine filters to remove yeast cells and any other impurities. Many small or microbreweries sell unpasteurized beer in bottles and cans. Nearly all beer distributed in kegs is also unpasteurized but is filtered to remove impurities. All unpasteurized beer products must be kept cold to ensure their highest quality. In still other cases, the brewer may elect not to filter or pasteurize the finished product.

Keg, or draft, beer is less expensive per ounce than bottled or canned beer, and contribution margins from sales are higher. However, spoilage can be a concern, adequate refrigeration space is required, and beer lines and taps must be frequently cleaned. There must be a demand that ensures sufficient sales of keg beer to justify the equipment and maintenance costs.

SPIRITS

Managers can select from many spirits available (*Exhibit 6.4*). They must determine the type of products such as gin, vodka, and whiskey, and also the quality level of each. The challenge is to balance customers' demands with an inventory of manageable size.

Every operation is different. The operation's concept is one important factor in determining the variety and type of spirits to offer. Additionally, the space available behind the bar, as well as in storage, should be taken in to account.

Exhibit 6.4

One important decision is to determine which spirits will be **well brands** (those served when there is no preference for a specific brand) and which will be **call brands** (a specific brand requested by a customer). The best call brands are sometimes referred to as "premium" or even "super premium," depending on product cost and selling price potential.

Specific well brands are selected because they cost less than better-known call brands. However, if the well brands are viewed as too "cheap," the establishment's reputation may suffer. Customers who do not specify a brand generally do not care what is served as long as the quality is acceptable.

In practice, a product that is neither the most nor least expensive is generally acceptable for well spirits. Spirits distributors can advise about well brands and premium products in keeping with the desired image of the operation.

Manager's Memo

Good managers select well and call brands based on what they know about customers' preferences and market conditions. Quality and value relative to the selling price are major factors. Other important factors include supplier prices, discounts, services, product reputation and availability, brands used by competitors, and suggestions from distributors and other knowledgeable sources.

Wise managers also know that customers' preferences can change. It is, therefore, important to obtain information from distributors about new products and how they are selling. Equipped with this information, managers may decide to offer other products on a trial basis and carefully monitor sales.

Receiving and Storing Beverages

Basic procedures for receiving alcoholic beverages are the same as those discussed for food in chapter 4. However, since most beverages are purchased by brand, the brand becomes the purchase specification against which receiving employees must check to ensure quality. This suggests that those receiving must only be able to read. However, even well-educated employees can have difficulty reading the labels of some wines, such as the German *Gewürztraminer Spätlese* and *Trittenheimer Apotheke Riesling Auslese*.

In addition to ensuring that the correct product was delivered, receiving employees must also confirm that the product is delivered in usable condition. Cases that appear to be wet or have been wet should be opened to confirm that there are no breakage or leakage problems.

After beverage products are received, they must be stored. Storage of alcohol products involves more than placing them in a storage area until needed. Most spirits are relatively nonperishable, and the shelf life of canned and bottled beer products is long. Wine is also easy to store under the proper temperature conditions. However, it is important to consider the quantity of products to be stored and how that storage process will be managed. There are several basic principles:

- Restrict access to the fewest possible employees. Beverages are theft-prone, and open-door policies that permit any staff member to enter beverage storage areas are unacceptable.
- Lock all beverage storage areas.
- Minimize the amount of product stored in bar areas. Behind-bar storage areas should be lockable and hold only the quantity of products needed for service during a shift.
    - Require that a manager be responsible for product issues to bartenders during a shift to encourage effective issuing and control access to storage areas.
    - Use a perpetual inventory system so managers will always know the quantity of products that should be available.

Exhibit 6.5

Managers must ensure that beverage storage areas are clean and well organized. Beverages other than keg beer should be kept off the floor and away from walls so walls can be easily cleaned (*Exhibit 6.5*).

Several storage principles will help maximize the quality of beers:

- Store keg beer between approximately 36°F and 38°F (2°C and 3°C).
- Store canned and bottled beer at 70°F (21°C) or lower, and rotate the stock when there is a delivery.
- Carefully monitor expiration dates on beer products.

Additional storage suggestions apply to spirits:

- Store spirits in clean, dry, and well-ventilated areas.
- Store sealed cases with the date of receipt marked on each case to allow for easy rotation.
- Store individual dated bottles upright on shelving to allow for easy rotation.
- Avoid excessive heat in beverage storage areas.

Wine storage is discussed later in this chapter.

Many managers keep a perpetual inventory on alcoholic beverages so they will always know the quantity of each beverage product available in inventory. This has several advantages:

- Reduced opportunities for theft
- The ability to determine how many additional products should be ordered
- Rapid and accurate counts of product usage

Managers should do a physical count to determine the quantity and costs of alcoholic beverages in inventory at least once per month. In many establishments, the inventory locations include central storage spaces and areas behind bars.

The manager, perhaps assisted by another manager or an entry-level employee without normal access to storage areas, counts the number of purchase units such as cases and bottles. The quantity on hand is multiplied by the cost per purchase unit to arrive at the inventory value.

Exhibit 6.6 shows a sample physical inventory form.

Exhibit 6.6

PHYSICAL INVENTORY FORM

Date of inventory: ______________________________

Taken by: ______________________________

Product	Purchase Unit	Units in Storage	Purchase Cost	Total Cost
(1)	(2)	(3)	(4)	(5)

Manager's Memo

Many small establishments do not have lockable refrigerator space to hold wines for service. One practical solution may be to purchase an enclosed and lockable rack that can be placed in a walk-in refrigerator or storeroom depending on the type of wine. Expensive and theft-prone items such as wine can be maintained under perpetual inventory in this lockable unit, and the products can be issued by a manager according to establishment policy.

With this plan, employees can obtain other products in refrigerated storage without having access to the locked items.

After an accurate inventory cost is established as shown in column 5 in *Exhibit 6.6*, managers can learn two things that are critical for beverage control: the inventory turnover rate and the current value of the beverage inventory. This information is needed for the monthly cost of goods (beverages) sold calculations (see chapter 4).

In addition to alcoholic beverages, other items must be stored for beverage-production use including drink mixes, juices, garnishes, and disposable products such as straws, napkins, and stir sticks. There must also be an adequate supply of all glassware.

Refrigerated storage may be necessary for cream, drink mixes such as for Bloody Marys and exotic drinks made with fruit juices, fruit and vegetables for garnishes, and fruit for drinks such as daiquiris. These items will likely be prepped before the beginning of each shift, so no long-term storage is needed. Ice cream and frozen mixes may require specialized storage space. Counter space for blenders, point-of-sale (POS) system, and a wide range of other equipment and supply items generally is necessary.

Local sanitation codes often mandate a three-compartment sink for glass washing. Some bar units have undercounter dish machines, but these are often inadequate for high-volume operations. In other establishments, bar glasses are washed and sanitized in the kitchen's dish machine and returned to the bar after cleaning. In many operations, counter, undercounter, and floor space is very limited.

Issuing Beverages

The proper issuing of and accounting for beverage products is just as important as their purchase and storage. Few operations employ full-time storeroom clerks to maintain beverage storage areas and issue beverages. As a result, control when issuing can be a challenge.

Bars should ideally be stocked on a regular schedule. For example, the bar may be set up for the midday shift in the morning and for the evening shift in the afternoon. When reasonable behind-bar par levels are established, they can help reduce the number of "emergency" issues that must be made during busy bar shifts. **Behind-bar par levels** are the number of bottles of each item that should be behind the bar at any time. During a shift, the par should represent the total number of full, partially full, and empty bottles.

Two important steps are required when issuing beverage products:

- Bottles should be issued on a bottle-for-bottle basis (exchange an empty bottle for a full bottle) to reestablish behind-bar par levels.
- An issue requisition should be completed by the bartender or manager who stocks the bar.

Exhibit 6.7 shows a sample beverage issue requisition form.

Exhibit 6.7

BEVERAGE ISSUE REQUISITION FORM

Shift ____________________ **Date:** ____________________

Beverage outlet: ____________________ **Completed by:** ____________________

Product	Number of Bottles	Bottle Size	Bottle Cost	Total Cost
(1)	(2)	(3)	(4)	(5)
Bar Gin	2	750 ml	$ 8.75	$17.50
Bar Rum	2	750 ml	7.75	15.50
Old Hoshler Whiskey	1	750 ml	12.50	12.50
Totals	**17**	______	______	**$173.50**

Issued by: ____________________ **Authorized by:** ____________________

Received by: ____________________

A bartender might assemble all empty bottles at the end of the shift. The types of spirits bottles emptied are noted in column 1, the number of empty bottles of each type is noted in column 2, and the size of the bottle is noted in column 3. The bottle cost is noted in column 4. This is easy to determine if the cost is written on the label at the time of issuing.

The total cost of each beverage product is determined (column 2 × column 4) and noted in column 5. The manager then signs the completed issue requisition, and the issue requisition form and empty bottles are taken to the central beverage storeroom. There, full bottles are issued on a bottle-by-bottle basis to replenish the bar par stock.

MANAGER'S MATH

The bartender has emptied the following bottles during a very busy shift. Please complete a beverage issue requisition form by copying the model:

Product	Number of Bottles	Bottle Size	Bottle Cost	Total Cost
Wee Dram Liquor	1	750 ml	$ 29.80	______
Bar Vodka	2	750 ml	16.40	______
Bar Tequila	1	750 ml	18.30	______
California Club	2	750 ml	21.65	______

What is the total cost of alcoholic beverages required to replenish the bar par?

(*Answer: Total cost is $124.20*)

BEVERAGE-PRODUCTION STANDARDS

Care is needed to ensure that beverages are prepared in an efficient and consistent manner. This requires a well-designed beverage-production area and a standardized process to produce drinks in a way that maximizes customer satisfaction.

Well-Planned Layout

The layout of a beverage-production area impacts its efficiency. Bars are often part of the customers' environment and must be strategically placed (*Exhibit 6.8*). The layout also affects employee productivity and the speed with which beverages can be produced.

As with other aspects of the physical facility, a variety of factors will impact the the layout. Factors including the operation's concept and budget will, of course, play some role in the final design. Manager's must take in to account all competing factors.

Exhibit 6.8

A beverage service and production area must promote efficient workflows so employees can do their jobs effectively. Layouts are influenced by the amount of space available. Other factors include the area's shape, the locations of entryways and exits, and the relationship of the bar to other locations including the kitchen, dining room, and restrooms.

Effectively designed layouts are important. An island bar in the center of the customer area may be just right for an establishment seeking to attract young professionals who want to mingle. However, this is not likely to be an acceptable bar location in an establishment that wants to attract families.

Bartenders need properly designed work stations so they can quickly perform their work without leaving the area when performing tasks such as obtaining products from remote storage rooms or retrieving clean glassware. Likewise, servers must be able to quickly order and pick up drinks and maintain service areas in a safe and sanitary manner.

> **OPEN FOR BUSINESS — KEEPING IT SAFE**
>
> Behind-bar work areas can be dangerous for bartenders because of ice and liquids spilled on the floor, especially during times of high production. While safety is always a priority, it is challenging to stop beverage production when spills occur during peak times. Slip-resistant flooring behind the bar makes good sense, as does a requirement that bartenders wear slip-resistant shoes. Appropriate flooring behind bars can have another advantage: reduced glass breakage.
>
> Also, train bartenders to always use a plastic or metal ice scoop when filling glasses. They should never use the glass itself or any other device made of glass. When a glass is broken in ice, the entire ice bin must be carefully cleaned to remove all glass fragments including small slivers. This cannot be done during a busy shift, and bartenders and customers will be inconvenienced while more ice and alternate storage are found.

Standardized Production Procedures

Controlling how alcoholic beverages are prepared is important for several reasons:

- Product costs must be controlled.
- Drinks prepared incorrectly will not meet standards and will disappoint customers.
- The quantity of alcohol served to each customer must be known if the establishment is to serve alcoholic beverages responsibly.

Managers must standardize drink production, and this requires careful monitoring of bar recipes and drink sizes. Beverage glasses in which drinks are served and even the ice used becomes important.

STANDARDIZED RECIPES

A bartender is similar to a chef because both, even if they are experienced, must follow standardized recipes. While experienced bartenders do not have to review a standardized recipe before they prepare a common drink such as Scotch and water, they do need to use the ingredients, quantities, and garnish that the recipe requires.

However, they may need to carefully study a recipe to prepare a less common drink such as a salty dog. The use of standardized recipes helps ensure that customers receive exactly what was requested and the establishment provides exactly what it intended.

PORTION SIZE STANDARDS

Standardized recipes for drinks, like those for food, should provide all information needed to properly prepare and serve the drink. However, special attention must be given to portion size, especially the amount of alcohol in the drink.

A standardized recipe for a gin and tonic may indicate two parts tonic water to one part gin (1 ounce of gin and 2 ounces of tonic water). If the manager calculates the beverage cost using this ratio, the customer will find the drink acceptable and beverage costs will be in line with expectations as long as the drink actually contains these amounts.

The actual beverage cost will, however, vary widely depending on the quantity of alcohol used in the drink. *Exhibit 6.9* shows the number of drinks and their costs and revenue per bottle when the drink size varies. Notice the increase in the alcohol cost percentage (Cost of alcohol per drink ÷ Selling price) as the quantity of alcohol in the drink increases.

Exhibit 6.9

BEVERAGE COSTS AT VARIOUS DRINK SIZES: LITER BOTTLE COST OF $15.00

Drink Size (1)	Number of Drinks per Liter Bottle (2)	Cost of Alcohol per Drink* (3)	Selling Price (4)	Total Revenue** (5)	Alcohol Cost Percent*** (6)
1.0 oz	33.8	$0.444	$5.00	$169.00	8.88%
1 1/4 oz	27.0	0.556	5.00	135.00	11.11
1 1/2 oz	22.5	0.667	5.00	112.50	13.33
1 3/4 oz	19.3	0.778	5.00	96.50	15.54
2.0 oz	16.9	0.888	5.00	84.50	17.75

*$15.00 (Bottle cost) ÷ Number of drinks per bottle (column 2).
**Number of drinks per bottle (column 2) × selling price ($5.00 in column 4).
***Cost of alcohol per drink (column 3) ÷ selling price ($5.00 in column 4).

THINK ABOUT IT . . .

Are there advantages to free pouring alcoholic beverages? If so, what are they? What are advantages to the use of portion tools for pouring? Which method would you require if you were a manager? Why?

Managers who allow bartenders to **free pour** alcohol (pour without a portioning tool) must ensure that bartenders neither **underpour** (use less alcohol than required by the recipe) nor **overpour** (use more alcohol than required by the recipe).

OTHER STANDARDS

Managers can select from many different styles of glasses when they plan how drinks should be served, and this is an important concern. For example, champagne served in a frosted beer mug will produce a poor-tasting glass of champagne and an unhappy guest.

Exhibit 6.10 shows examples of glassware used in many establishments.

Exhibit 6.10

COMMON TYPES OF GLASSWARE

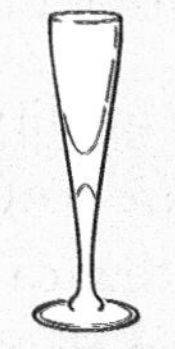 Champagne Flute Used for Champagne	**Champagne Flute (6 to 8 ounces):** This glassware's long stem can be grasped with the fingers to keep the drinker's hand from transferring body heat to the cold contents.	 Collins Glass Used for Larger Mixed Drinks	**Collins Glass (6 to 16 ounces):** This popular shaped all-purpose glass is used for mixed drinks, water, soft drinks, juice, or other nonalcoholic beverages.
 Wine Glasses Used for Wine	**Wine Glass (4 to 12 ounces):** Red wines are typically served in larger glasses to allow the wine's bouquet (aroma) to be enjoyed. White wines are generally served in smaller glasses so they can retain the wine's cooler serving temperatures. Dessert wine glasses may be even smaller.	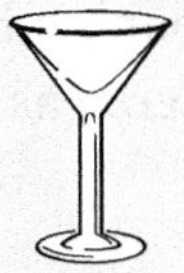 Martini Glass Used for Martinis	**Martini Glass (4 to 6 ounces):** The classic martini glass has straight edges expanding outward from the stem at approximately a 60-degree angle.
 Snifter Used for Brandy	**Snifter (16 to 18 ounces):** This glass is popular for brandy or Cognac. The large bowl maximizes the surface area of the beverage and allows the drink to be gently swirled to release the maximum amount of bouquet.	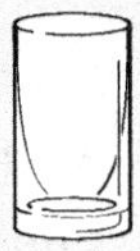 Highball Glass Used for Mixed Drinks	**Highball Glass (8 to 10 ounces):** The highball glass is tall and slender and displays colorful drinks very well.
 Irish Coffee Glass Used for Coffee or Hot Alcoholic Drinks	**Irish Coffee Glass (6 to 16 ounces):** This tempered clear glass mug attractively displays hot drinks including Irish coffee, regular coffee, cider, tea, or other served-hot beverages.	 Rocks Glass Used for Smaller Mixed Drinks; Shots of Alcohol Over Ice	**Rocks Glass (8 to 10 ounces):** A rocks glass is short, heavy, and wide-rimmed and is essential for any bar.
 Beer Pilsner or Mug Used for Beer	**Beer Pilsner or Mug (12 to 16 ounces):** Many managers view the pilsner-style glass as the best for beer. It has a narrow, footed base and expands outward toward the top.	 Shot Glass Used for Shots of Alcohol	**Shot Glass (2 to 2.5 ounces):** This glass is used for many types of spirits that are served "straight up" (without ice).

Other types of glasses include cordial glasses used to serve liqueurs, dessert wines, and sweet, flavored spirits and other specialty glasses used for hurricanes, piña coladas, mai tais, and frozen and on-the-rocks daiquiris.

Ice used for beverage production can have a large impact on the quality of the beverage. Large ice cubes leave space between them when they are scooped into a glass. This permits a larger amount of mixer to be added that can dilute the drink more than intended. Smaller cubes or shaved ice pack a glass, and a smaller volume of mixer can be added. This may give the impression of a stronger drink.

Can an operation use one ice machine for all uses such as drink production, salad bar, and food production? The answer to this question must be carefully considered as beverage recipes are developed and standardized.

MANUAL AND AUTOMATED BEVERAGE PRODUCTION

All alcoholic beverages can be produced manually. Some drinks such as cocktails made with few additional ingredients, and highballs made of a spirit and a larger quantity of a mixer, can also be automatically produced and dispensed.

Exhibit 6.11

Manual Beverage Production

All bartenders must be properly trained because they must know how to prepare many drinks regardless of the system used. Bartenders must follow the property's recipes whenever they prepare a drink, and a measuring tool such as a jigger should always be used to ensure portion control. A **jigger** is a small shot glass–type tool used to measure the amount of alcohol in drink preparation (*Exhibit 6.11*).

Many jiggers have lines that indicate the quantity of spirits in the same way that food measuring tools provide quantity indications. When producing a drink, the bartender holds the jigger and pours until the alcoholic beverage reaches the line. This is done in sight of the customer if seated at the bar. Then the jigger's contents are poured into the proper glassware. This method is accurate and helps ensure that beverage costs are in line with plans and that drinks will be of consistent quality. It also assists in the property's efforts to serve alcoholic beverages responsibly because a count of the drinks billed to the customer indicates the total quantity of alcohol consumed.

THINK ABOUT IT . . .

Some customers think that bartenders using jiggers are going out of their way to ensure no extra alcohol is served and, therefore, value is diminished.

What do you think about using a measured pour to produce drinks?

Some managers require bartenders to use measured pour spouts to produce cocktails and highballs. A **measured pour spout** controls the beverage quantity by only allowing a specified amount of alcohol to be dispensed. For example, if a manager decides to use a 1-ounce portion, after inserting the spout in the bottle, only this amount can be poured without inverting the bottle to pour a second portion.

Pour spouts are available at a variety of prices. Features range from simple portion control to automated beverage-production systems that calculate product usage and compute expected revenue levels and product cost percentages.

Automated Beverage Production

Many managers of high-revenue-volume operations automate some of their beverage-production processes. Highballs and cocktails can be dispensed in a predetermined amount by a machine activated by the bartender. In the best systems, the dispensing equipment is connected to the bar's POS system. This helps ensure that each drink produced is automatically charged to the proper customer.

This setup also reduces opportunities for bartenders to give away free drinks. A record is kept of the number of drinks produced, the amount of alcohol of each type dispensed, and the revenue value of the drinks. There are other possible benefits of an automated beverage dispensing system:

- The elimination of over- or underpouring
- Reduced spillage
- Reduced drink pricing errors
- Accurate record keeping for all products sold
- Fewer bartender errors
- Less need for bartender supervision
- Lower and more consistent product costs
- Reduced cost per product ounce when purchasing in larger containers
- Reduced legal liability potential because the amount of alcohol is controlled and recorded

Advances in automatic dispensing equipment have eliminated many earlier difficulties with these systems. Equipment breakdowns and malfunctions have been greatly reduced, as have the costs for system maintenance and upkeep. Managers of establishments with high-volume sales should investigate potential advantages and disadvantages of automating part of their beverage production.

BEVERAGE SERVICE METHODS

After preparation, drinks must be served to customers. Three basic methods can be used.

Service by Bartenders

Customers can place their orders directly with the bartender, who then produces the drink and serves it (*Exhibit 6.12*). This is a labor-efficient delivery system for drinks served at the bar, and it may also be useful for lounge service when there are few customers.

Exhibit 6.12

Bartenders must have ample space to work efficiently. Often, the physical length of the **public bar** (a bar at which customers can be seated) is determined by the need for a specific number of seats or standing spaces in front of the bar. The number of drinks that can be effectively served to customers with this service style is basically determined by the design of the work areas behind the bar and its total length. In addition, the number of bartenders that can effectively work behind the bar is determined in large measure by its length.

Service by Beverage Servers

In some operations, beverage servers take customer orders, a bartender prepares them, and the servers deliver them. This service style is most common in dining situations and in many upscale bar or lounge settings.

When used, the area in which orders are placed and drinks are picked up should be convenient for server access. As the distance of the bar from customer seating areas increases, service time also increases. Also, the section of the bar where servers pick up drinks must be of adequate size to hold prepared drinks until they are picked up. In some operations, servers garnish their drinks, and space will be required for garnish storage if this is done.

Additional space may also be needed to store service items such as additional glasses, ice, garnishes, and napkins. A potential problem is that customers may stand or otherwise block server access to the beverage pickup counter. The design of the bar and railings or other devices can help keep customers away from these areas.

Service by Bartenders and Servers

In some operations, the bartender may serve customers at the bar and a server may serve drinks in the lounge area. This system is often used in establishments where guests can consume meals at the bar. Larger establishments with designated lounge and dining areas also commonly use this type of service. During slow periods, there may not be a beverage server present. Then the bartender serves drinks to customers at tables as well as at the bar.

Bartenders should have easy access from the bar to customer service areas. Regardless of the service style selected, bar areas must be efficiently designed and properly stocked to ensure maximum bartender and server productivity.

ENHANCING WINE SALES

Selecting wines is among the most difficult beverage purchasing decisions made by many managers. There are thousands of wineries around the world, and each may make several types of wine.

Regardless of the food served in an operation, there will be wines available to complement menu offerings. Remember that when wine is served with a meal, it can elevate the dining experience to a special occasion and that will ultimately result in increased customer satisfaction and greater profits.

The proper purchasing of wine requires significant knowledge about the product, an understanding of value, and a feeling for what customers will select. Some managers hire wine consultants to assist with wine selection when an extensive wine cellar is desired. However, good managers are able to select an assortment of useful wines for a more modest inventory.

THINK ABOUT IT . . .

The customer's preference for any wine with any food at any time during the meal is the best "rule" to follow. Why do you think cooking professionals and other culinary experts support food and wine pairings?

Wine and Food Pairings

The wines available should complement the menu items and sell at prices consistent with those for food. There is a long tradition of **wine and food pairings**: the idea that some wines go better with some food items than others, and that wine should be selected after the food item to match the food. However, the preferences of customers are always most important.

Some food menus offer wine suggestions with entrées, but this can present challenges. Consider a couple, one of whom has ordered a beef dish and the other a chicken item. The menu may suggest only one wine for each dish: a red wine for the beef and a white wine for the chicken. If the wines are available only by the bottle, it is unlikely that both of the suggested wines will be selected. Servers should be trained to make suggestions about wines available by the glass that will complement each entrée.

Exhibit 6.13 contains general wine selection information that can assist in making appropriate purchasing decisions.

Exhibit 6.13

WINE AND FOOD PAIRINGS

Wine Type	Serve With
Red Wines	
Cabernets and Red Bordeaux	Lamb roasts and lamb chops, all cuts of beef steak, roast duck, goose
Merlot	Beef and lamb roasts, venison, sirloin steaks, grilled or roast chicken
Pinot Noir	Roast chicken, rabbit, duck, grilled salmon, grilled tuna
Shiraz	Grilled or roast beef, game meats, BBQ, pizza
Sangiovese	Roast pork, roast chicken, pasta, grilled vegetables, Italian sausages, pizza
White Zinfandel/White Merlot	Seafood salads, pastas, grilled chicken, grilled pork loin, Mexican food
White Wines	
Chardonnay	Sole, halibut, cod, scallops, lobster, roast chicken, pasta with seafood or chicken
White Riesling	Roasted pork, chicken, veal, smoked salmon, paté of all types, sushi
Sauvignon Blanc	Fish, shrimp, calamari, fresh oysters, sashimi
Pinot Grigio	Pastas, grilled chicken and shrimp, veal
Sparkling (Champagne)	Caviar, fresh oysters, sushi, sashimi, lobster

Wine Lists

A good wine list offers a reasonable selection of wines that go well with the food offered at prices that encourage customer purchase. A **wine list** is a special menu that identifies the wine selections offered along with their selling prices.

Many principles used to design food menus also apply to planning wine lists. Other suggestions include the following:

- Remember food and wine pairings. An establishment offering a variety of menu items may offer numerous types of wines from many areas. For instance, an Italian establishment may emphasize Italian wines with a wide selection of varieties.
- Match wine prices to the selling prices of food. Few managers would offer high-priced wines when food items are inexpensive, and the reverse is also true. Some managers choose to offer a few higher-priced wines for customers who know about and prefer a high quality in wines.
- Make it easy for customers to order. Many establishments list wines by bin number, and then a customer can order "number 201." A **bin number** tells the location in a wine storage area where a specific wine is stored.
- Offer wine by the glass so customers ordering different entrées can enjoy a wine that complements their food. Also, consumers limiting their alcohol consumption might purchase a glass rather than a bottle.
- Include wine descriptions such as origin, flavor, and special characteristics.
- Avoid repetition. For example, it is not typically necessary to offer several merlot wines on a short wine list.
- Be sure information is accurate and correctly spelled. Obtain the information from the bottle label and ask distributors about region, vineyard, and vintage. **Vintage** wines are grown from grapes in one vineyard during one season.

More detailed menu item–specific information must also be considered when wine lists are developed:

- Include light and dry red wines when the menu features beef.
- Include full-bodied dry red wines when the menu features game and game birds.
- Include dry white wines when the menu features poultry, pork, veal, or seafood.
- Include semi-sweet or sweet wines to complement desserts and at any time that customers prefer them.

Manager's Memo

Wines can be listed on wine lists in several ways:

- In the order they would be consumed: appetizer wines, entrée wines, and dessert wines.
- By type, such as color (red wines followed by white wines) or region (French wines followed by domestic wines). Other examples include type of grape, vintage, or degree of dryness or sweetness.
- By price range, such as least to most expensive.

Some wine lists use more than one organization scheme. For example, wines may be listed by color and, within the color category, be listed from most to least expensive.

- Include a sparkling (carbonated) wine to be sold with dessert or at any other time customers want to order it.
- Select wines that are readily available in quantities consistent with the establishment's anticipated sales volumes.

Wine Pricing

Product cost is often a factor in wine pricing. Many customers are aware of a favorite wine's retail selling price and may question an establishment's selling price when it is several times higher.

Wine list developers should recognize the need for selling prices to encourage sales while generating a profit:

- Keep selling prices in line with the operation's overall pricing structure.
- Offer some inexpensive wines for cost-conscious customers.
- Think about customer choices. Customers desiring an alcoholic beverage typically have choices including beer and other alcoholic drinks. How does the selling price of a glass or bottle of wine relate to that of other available alcoholic beverages?
- Consider contribution margin (Selling price − Product cost) instead of cost percentage (Product cost ÷ Selling price) when setting the selling price. The contribution margin is more important because it represents the money that remains to pay for other expenses and contribute to profit after the product has been paid for.

Manager's Memo

A markup factor is the number by which a product's cost is multiplied to arrive at a selling price. It is often used to price wines. Here are suggestions:

Type of Wine	Markup Factor
Bulk wines (box or house wines)	4–5
Inexpensive bottled wines	3
Moderately expensive bottled wines	2
Highest-price bottled wines	Less than 2

Wise managers know that suggestions are just that: suggestions. They understand their customers' definitions of value. Then they price their wines and all other products at a level that will encourage sales because of the value the customers will receive.

Wine Storage

Storage concerns must be addressed to keep wines at their highest quality until their sale. General storage suggestions include the following:

- Store bottled wine on its side.
- Store red wines at temperatures between 50°F and 70°F (10°C to 21°C). If at all possible, store them at **cellar temperature**: a constant storage temperature between 55°F and 60°F (13°C to 16°C).
- Store white and sparkling wines in refrigerators if they are to be used within a few months of purchase, or between 50°F and 70°F (10°C to 21.1°C) if they will be held longer.
- Avoid excessive light, humidity, and heat in wine storage areas, as these can damage wines and their containers.

Wine Service

Establishments with high-volume wine sales may employ a **sommelier**: a service employee with extensive knowledge about wine including wine storage and wine and food affinities. The sommelier advises customers about wine selection, takes wine orders, and presents and serves the wines selected. Most operations do not have this specialized assistance available, and the food or beverage server is responsible for assisting customers with wine selection.

An important first step in selling wine is to have some product knowledge about the wines that are available. Servers should also know how to properly pronounce each wine's name.

In some establishments, tables are set with wine glasses. The setting may include a glass for white wine and a separate glass for red wine. If a wine is not ordered, both glasses should be removed from the table. If one type of wine is requested, the glass that is not needed is removed. In other establishments, wine glasses are not set on the table before the customers arrive. Instead, they are brought to the table if wine is ordered.

Exhibit 6.14 reviews steps that form the basis for professional wine service.

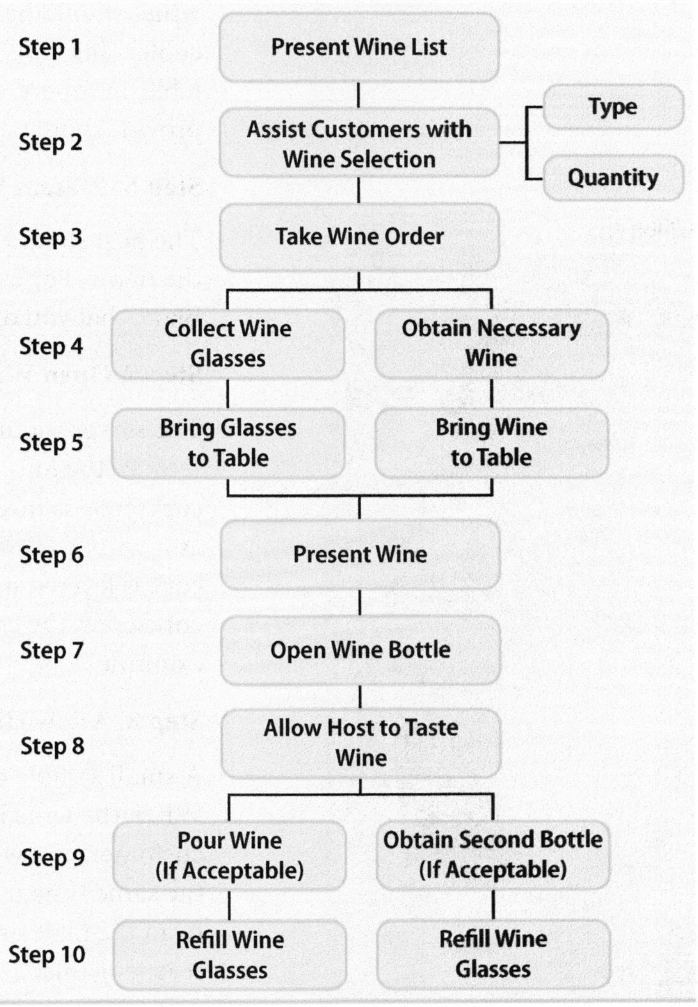

Exhibit 6.14

STEPS IN PROFESSIONAL WINE SERVICE

Exhibit 6.14 outlines 10 steps:

Step 1. Present Wine List

Wine lists, like food menus, represent the operation's brand. They should be clean and neat because they are significant selling tools. They should be brought to the table for presentation to the customer with a comment such as "I am proud to present our wine list." This can be followed with the suggestion "If you have any questions about our wines, I will be happy to answer them. Otherwise, please take a few moments to look over the list." In some establishments, wine lists are placed on the table when it is set for service.

Step 2. Assist Customers with Wine Selection

Some customers are likely to be very knowledgeable about wine and will not desire or appreciate assistance. Others may have questions about the type to select relative to the food being ordered or the quantity to order. There are approximately five servings of 5 ounces each per bottle; a full bottle holds 25.6 ounces. Therefore, a full bottle will likely be sufficient for each two to three people depending on portion size.

Step 3. Take Wine Order

The wine order should be taken and repeated to ensure that there is no communication problem.

Step 4. Collect Wine Glasses and Obtain Necessary Wine

Some establishments use the same wine glass regardless of the wine type selected; others use specific glasses for specific types. Establishments use various systems for servers to obtain the wine ordered. White wines are typically served chilled; red wines are typically served at approximately room temperature. These factors influence where the wine will be stored until it is issued to service staff.

Step 5. Bring Glasses and Wine to Table

If wine glasses are not preset, they should be brought to the table on a small serving tray and should be handled by their stems, not by their bowls or rims. They should be placed to the customer's right. Red wines can be brought to the table on a tray or in a wine basket. Chilled white wines should be brought to the table on a small serving tray or in a wine cooler (bucket). In some establishments, wine coolers are placed on the table; in others, a cooler stand is placed on the floor to the host's right to provide more table room.

Step 6. Present Wine to Host

The host should be allowed to read the label while the server pronounces the name. For example: "This is a bottle of the Robert Mondavi Fume Blanc that you ordered."

Exhibit 6.15

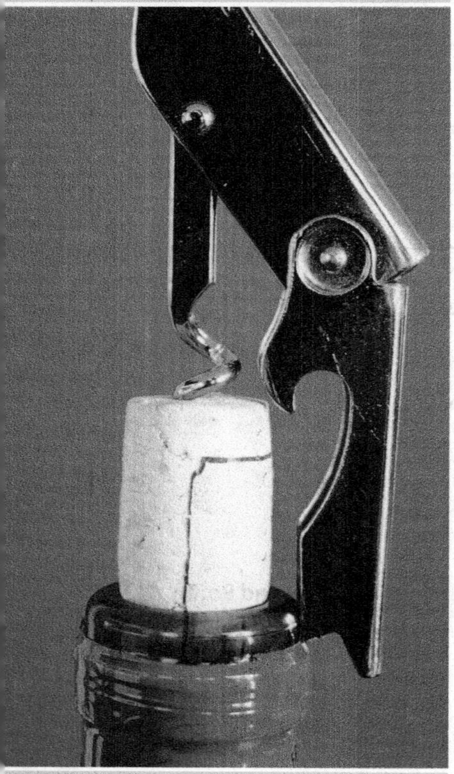

Step 7. Open Wine after Host Approval

The server should use the knife blade attachment on a corkscrew to remove the foil, which is placed in the server's pocket or apron. The corkscrew is inserted into the center of the cork and is screwed in as far as it will go (*Exhibit 6.15*). The lever is placed on the bottle's lip, and the cork is levered up until it can be removed. Once removed from the corkscrew, the cork can be presented to the host, who may want to examine it.

Step 8. Allow Host to Taste

A small sample of the wine is poured into the host's glass for sampling. When the wine is poured, the bottle's label should be facing the customer, and when pouring is complete, the bottle should be twisted at the same time it is tipped up to reduce drips. Glasses are not removed from the table as wine is poured unless, for example, the customer is seated against a wall or in a booth, or there is another reason that pouring in place is impractical.

Step 9. Pour Wine after Host Approval

Managers should establish the portion size to be poured in wine glasses, and this decision will be based in part on the shape and size of the wine glass.

Begin with the eldest woman, serve other women, then other guests, and finally the host. If the host does not approve the wine, the server should retrieve a second bottle. Improper processing, transporting, or storing conditions sometimes yield an unsatisfactory bottle of wine. Sometimes, however, the customer may just be unsatisfied with the wine's taste. Care should be taken that this is not caused by an improper description on the wine list or by the server. Alternatively, the server may suggest a different wine or request that the manager or other staff member with more extensive wine knowledge assist the customer. When a new bottle of wine is brought to the table, steps 6 through 9 should be repeated.

Step 10. Refill Wine Glasses

Servers should know when to refill glasses. When a guest's wine glass is almost empty, the server can inquire "Would you care for more wine?"

There is a great deal of tradition and "showmanship" in professional wine service. Proper use of these procedures is appreciated by many customers and is an important part of their enjoyment of the wine and of the meal itself.

SUMMARY

1. **Explain the importance of providing quality non-alcoholic beverages.**

   Managers should be concerned about the service of non-alcoholic beverages because they are popular and profitable. Attention to simple but important details can help ensure that water, coffee, tea, soft drinks, and milk consistently meet the establishment's quality standards.

2. **Describe federal, state, and local regulations that impact the purchase and management of alcoholic beverages.**

   The federal Bureau of Alcohol, Tobacco, Firearms and Explosives (ATF) enforces federal laws related to the manufacture and sale of alcohol. State licenses permit operations to sell beer only; beer and wine; or beer, wine, and spirits. Managers must know and comply with all regulations applicable to what is sold; where, when, how, and to whom; and in what quantity.

   Many municipalities have their own liquor authority and regulations that affect the sale of alcoholic beverages, and local laws are often stricter than the applicable state laws. Local regulations also deal with food and fire safety code regulations that impact beverage operations. The storage of alcoholic beverages and supplies used to prepare them are of concern during food safety inspections.

3. **Explain basic management practices applicable to selecting, purchasing, receiving, storing, and issuing alcoholic beverages.**

   In control states, the state is the only supplier of liquor. In license states, the state grants licenses to wholesalers, distributors, or manufacturers to sell alcohol.

   Many types of beers and liquors are available, and managers must determine those that are most popular. Product costs and selling price potentials are additional concerns.

   Storage concerns relate to maintaining quality and reducing theft. Physical inventories on at least a monthly basis are necessary to determine the quantity and cost of products in inventory. Beverage issues should be on a full-bottle-for-empty-bottle basis to reestablish the behind-bar par level.

4. **Explain beverage-production standards for bar layout and drink preparation procedures.**

   A beverage-production area (bar) must be designed for employee efficiency. Control of alcoholic beverages is important to manage costs, meet standards, and ensure that alcohol is served responsibly. Drink production is standardized using standardized recipes that specify portion size and by ensuring that quantity measures are always used. Glasses and even the ice used have an impact on beverage quality.

5. **Describe the use of manual and automated beverage-production methods.**

   Even experienced bartenders must follow standardized recipes for beverages. Spirits should be measured with a portion control tool (jigger) or with a metered pour spout.

   High-revenue-volume operations may automate production of highballs and cocktails using dispensing systems that automatically determine the customer charge, assess the quantity of liquor used, and ensure that the correct portion of alcohol is served.

6. **Describe three beverage service methods.**

   Three basic methods can be used to serve alcoholic beverages to customers. First, bartenders serve drinks to customers at the bar and may serve customers in nearby areas during slow times. Second, beverage servers serve customers in the bar lounge or other areas. Third, bartenders and servers are used. With this system, the bartender may also serve food at bar areas.

7. **Explain procedures for managing wine sales.**

   Effective wine purchasing requires product knowledge and an understanding of value and customer preferences. There is a long tradition of food and wine pairings in which wines are selected after the food item to complement the food.

   A wine list serves the same purpose as a food menu. There are many strategies to ensure customers have desired wine alternatives. The selling price should be in line with the operation's pricing structure and generally based on marking up the cost.

Managers must plan for wine storage and monitor the temperature, light level, and relative humidity of storage areas unless wines will be used soon after purchase.

Wines sold by the bottle should be presented and served according to procedures established by the property. These typically allow the host to sample wine and make wine service an important part of the dining experience.

APPLICATION EXERCISE

There are many methods dishonest bartenders use to steal beverage revenue from the operation. It is the manager's responsibility to design and manage beverage revenue systems in a way that minimizes this temptation.

Enter "common bartender theft methods" into a search engine. Use the results to make a list of 10 methods. Explain what you as a manager would do to address these concerns with bar revenue control procedures.

REVIEW YOUR LEARNING

Select the best answer for each question.

1. **To ensure that only the highest quality of coffee is served, how many minutes can coffee be kept before service until it is discarded?**
   A. 20
   B. 30
   C. 40
   D. 50
2. **What process kills the microorganisms most likely to cause spoilage in beer?**
   A. Sterilization
   B. Fermentation
   C. Pasteurization
   D. Homogenization
3. **What is a liquor that customers do not order by a specific name (brand)?**
   A. Well brand
   B. Call brand
   C. Regular brand
   D. Premium brand
4. **At a minimum, how frequently should a physical inventory of alcoholic beverages be performed?**
   A. Once per shift
   B. Once per day
   C. Once per week
   D. Once per month
5. **Alcoholic beverages should be issued from storage in a quantity required to achieve what goal?**
   A. Hold overflow from central storage
   B. Meet sales estimated for all shifts
   C. Maintain behind-bar par levels
   D. Fill behind-bar storage areas
6. **What is an advantage of standardized recipes in bar operations?**
   A. They allow the establishment to control alcoholic beverage portions.
   B. They allow any available employee to prepare acceptable drinks.
   C. They ensure that customers receive a value for drink purchases.
   D. They allow the manager to purchase inexpensive liquors.

7. Which type of wine is typically served in the largest glass?

A. Red

B. White

C. Dessert

D. Sparkling

8. What is the type of bar at which customers can be seated?

A. Main bar

B. Public bar

C. Service bar

D. Cash bar

9. Wine and food pairings suggest which type of wine to be served with beef?

A. Red

B. White

C. Dessert

D. Sparkling

10. The number of bottles of each beverage that should be behind the bar at any time is called the behind-bar

A. purchase total.

B. safety level.

C. sales quota.

D. par level.

FIELD PROJECT

1. Make an appointment with a local wine distributor and ask him or her the following questions:

   A. What type of assistance can a wine distributor give to a restaurant or foodservice manager as a wine list is developed?

   B. What should well-informed food or beverage servers know about the wines offered by their employer?

C. What is the best way to train servers about the wines offered by an establishment?

D. What are several of the most popular Italian wines? Do customers' preferences change frequently?

7 Facilitating Performance of Production Staff

INSIDE THIS CHAPTER

- Develop Job Standards
- Use Job Descriptions
- Train Employees to Attain Job Standards
- Use Checklists
- Evaluate Employees against Job Standards
- Control Labor Costs during Production

CHAPTER LEARNING OBJECTIVES

After completing this chapter, you should be able to:

- List and describe the procedures for performing a position analysis to create job standards.
- Explain strategies for developing job descriptions.
- Describe procedures for implementing performance-based training programs that help employees consistently attain job standards.
- Explain how checklists can help managers ensure that job standards are being met.
- Describe procedures for evaluating employees' performance as needed to maintain job standards.
- Describe a five-step process for controlling labor costs during production.

KEY TERMS

action plan, p. 211

bona fide occupational qualifications (BFOQs), p. 199

competencies, p. 200

cost-effective, p. 200

essential function (Americans with Disabilities Act), p. 199

job description, p. 194

job specification, p. 196

on-the-job training, p. 201

operating budget, p. 212

overtime, p. 216

performance standard, p. 195

position analysis, p. 194

reasonable accommodation, p. 199

task, p. 194

task breakdown, p. 195

task list, p. 194

training lesson, p. 201

variance (budget), p. 216

CASE STUDY

"Look, I know we need to use standardized recipes to make food items," said Gilda, a cook at Green Gardens Grill. She was talking to Charlie, another cook.

"But," she continued, "I certainly don't think we need a recipe for how to do our job!"

"I know what you mean," replied Charlie. "Our new boss wants us to write down everything, like how we wash pots and pans and how we receive products, in almost the same way we have written out our recipes."

"I think we do a good job, and I don't see why we should write down things we have done for so long that they are automatic," said Gilda.

1. Do you think the manager did a good job of explaining the need for employees to write down the procedures they follow for the tasks they do? Why or why not?
2. What should the manager do on discovering that the employees do not want to develop job tasks and job breakdowns?

THINK ABOUT IT . . .

How does a cook know how to clean the kitchen at the end of a shift, and how does the manager know if it was done correctly?

As a manager, how would you decide the way that the kitchen should be cleaned?

DEVELOP JOB STANDARDS

Managers must do a great amount of planning before food or beverage production begins. Their activities include developing the menu; standardized recipes; procedures for purchasing, receiving, storing, and issuing products; and standards for food production. This chapter explores directing the work of food-production staff, and a first step is to develop the job standards that drive how the work should be done.

A **position analysis** can be used to identify each task an employee must do and explain how it should be done. A **task** is a duty or activity that is part of a job (position). For example, one task in the dish-washer's job is to wash dishes. Once developed, a position analysis gives managers operating procedures that can be used to evaluate how well work is being done.

Preparing a position analysis involves four basic steps:

1. Develop a task list.
2. Decide how tasks can be separated into small activities.
3. Think about performance standards.
4. Write a **job description** that indicates the most important tasks that are part of a job.

The first three activities will be examined in this section. The last step will be covered in the section that follows.

A **task list** specifies all tasks that are part of a job. It focuses on the activities an employee must be able to do. For example, a cook must know how to operate food-production equipment. A bartender may need to know how to restock the bar.

Managers can use basic procedures to develop a task list:

- Ask supervisors and experienced workers in the position questions such as "Describe what you do in a normal work shift, from when you begin work until you complete your shift."
- Review written information about a position, such as any existing job descriptions and training materials used to teach new employees.
- Observe staff members as they work. Compare what they actually do to what they said they did when asked (*Exhibit 7.1*).

Exhibit 7.1

After reviewing the information, a manager can develop a list of the tasks that make up a job. Similar tasks can be combined, and factors such as differences between work shifts or production volumes can be identified. Then managers will know the scope of training requirements for new employees. They will also have benchmarks for evaluating performance: Is the employee doing the work the correct way?

A **task breakdown** tells how each task in the task list should be done. A cook may need to "Properly operate the deep fryer." Developing the task breakdown for this activity may be simple, because the manager can probably use the operating instructions provided by the equipment manufacturer to divide this task into steps.

There are several benefits to task breakdowns:

- They tell the correct way to perform a task to best ensure that performance standards are met.
- Trainees benefit from copies of task breakdowns given to them because the breakdowns can be reviewed as needed.
- Trainees can practice each step and then compare what they did to the steps in the task breakdown.

Managers can develop task breakdowns by performing a series of steps:

- Watch an experienced staff member perform a task.
- Record each activity or step in sequence.
- Ask the staff member to review the information to confirm its accuracy.
- Share the information with other experienced staff members and their supervisors.
- Make modifications, if necessary, so everyone agrees with the work method.
- Review the task work sheet with the employee's supervisor and the employee.
- Confirm the final task breakdown by observing an experienced person who performs the task using the identified procedures.

Performance standards specify required quality and quantity outputs, which define the "correct" way to perform a task. Proper performance must be clearly defined so employees know what is expected and managers know when performance is acceptable. One common example of a quality output is a portion of food produced according to an establishment's standardized recipe. An example of a quantity output also comes from the same standardized recipe. If a cook was asked to prepare 50 servings of a menu item, he or she would use the applicable recipe. Since it specifies the procedures to be used, the recipe helps control the time needed to produce the required quantity of portions.

Performance standards should be reasonable. They should be challenging but achievable. Employees should learn the procedures identified by the task breakdowns, and they must have the tools and equipment needed to work correctly.

THINK ABOUT IT . . .

A standardized recipe is a task breakdown. A cook must know how to prepare entrées. Standardized recipes explain every detail of preparation.

Do most managers consider standardized recipes in this way? Why or why not?

Performance standards must be specific so that they can be measured. Compare these statements: "The cook will know how to prepare twice-baked potatoes" or "The cook will be able to prepare twice-baked potatoes according to the standardized recipe." The second standard is better because it can be objectively measured.

Exhibit 7.2 shows a sample task list for a bartender and also a task breakdown for one task on the task list.

Exhibit 7.2

SAMPLE TASK LIST AND TASK BREAKDOWN FOR BARTENDER

Part A: Sample Task List

- Inspects the bar before opening to ensure adequate supplies are available.
- Follows setup procedures.
- Mixes, prepares, and serves drinks for customers.
- Collects payments for drinks served.
- Maintains and cleans bar area and equipment.
- Carefully follows all laws and establishment's policies and procedures regarding alcoholic beverage service and stops service to intoxicated guests.
- Carefully follows closing checklist at end of shift.

Part B: Sample Task Breakdown for Stopping Service to Intoxicated Guests

- Alert a backup.
- Enlist the help of other guests if possible.
- Wait until the guest orders the next round before stopping service.
- Inform the guest that alcoholic beverage service is being stopped.
- Offer nonalcoholic alternatives.

USE JOB DESCRIPTIONS

The last step in position analysis is to summarize the major tasks in the position. The document used for this purpose is a job description.

Uses of Job Descriptions

There are several reasons to use job descriptions. First, they tell exactly what employees must do in their position. They also provide an important foundation for many other functions in the operation:

- **Recruiting and Screening:** Job descriptions often provide job specification information in the form of a list of the personal qualifications necessary for an employee to be successful in the position. When this information is included, potential job candidates can screen themselves before applying. It provides a foundation for evaluating job applicants and developing legally defensible interview questions and screening practices.

- **Selection and Orientation:** The job description clarifies what the job involves and provides direction for selection and orientation programs.
- **Professional Development and Training:** Managers often review job descriptions as they plan and evaluate professional development and training programs, as shown in *Exhibit 7.3*. This helps them determine how an individual's performance compares to work expectations and whether more training may be needed. In addition, the use of job descriptions makes it easier to identify how to develop employees for other positions.

Exhibit 7.3

- **Performance Appraisal Programs:** A job description provides basic information and regular expectations for a position. This makes it useful for performance evaluation including the creation of employee improvement programs.
- **Salary Administration:** Job descriptions and job specifications help in developing compensation or salary ranges for positions. They provide a starting point for classifying jobs into categories that require common knowledge and skills. For example, all culinary positions would be in a culinary class. Job descriptions also make it easier to compare positions between operations. For example, the position dining-room manager can include very different responsibilities in different operations. Even small operations can use job descriptions informally to help determine if their compensation is competitive.
- **Safety and Security:** Job descriptions can define responsibilities for ensuring food safety and security and preventing injuries. For example, a cook position might include duties such as "Handle, prepare, and store food following Hazard Analysis Critical Control Point (HACCP) guidelines." A manager position might include security duties at closing such as "Ensure the establishment is locked and the alarm is set."
- **Union Relations:** Job descriptions are often carefully studied when misunderstandings occur between a union employee and supervisors or when there are union grievances. They also clarify the scope of a job and the range of responsibilities an employee is expected to undertake, which can be a factor during contract negotiations.
- **Legal Proceedings:** Job descriptions may be reviewed in lawsuits, administrative hearings for discrimination claims, or other legal proceedings. Well-written job descriptions provide more protection to an operation.

THINK ABOUT IT . . .

Some job descriptions include job specifications outlining the qualifications candidates should have. These may include education, experience, and skills. Attitudes, especially for dining-room positions, may also be specified.

How could knowing job specifications help an applicant?

Developing Job Descriptions

A well-written job description includes the information needed to understand the job. It should be specific enough to tell what the job involves, but not so detailed that it defines minor job tasks that may change. Job descriptions that are too detailed can also discourage teamwork because they can suggest "who does what" in a way that assigns specific responsibilities to only one position.

Exhibit 7.4 shows a sample job description that includes job specification information. Typically, an operation uses the same format in each job description and provides an overview of the same categories of job information. Some establishments may use one job description format for management and another for nonmanagement positions.

Exhibit 7.4

SAMPLE JOB DESCRIPTION

Kitchen Manager

Part I: Reports to: Restaurant Manager

Part II: Overview of Position: Supervises all kitchen-related operations. Selects, discharges, supervises, trains, and evaluates all food production employees to ensure quality and cleanliness standards are consistently attained. Purchases/issues food items and supplies. Serves on the menu planning team and manages product costs according to budget requirements.

Part III: Job Tasks

1. Supervises all kitchen-related activities.
2. Selects, discharges, supervises, trains, and evaluates food preparation personnel to ensure that quality and cleanliness standards are consistently attained.
3. Purchases/issues food items and supplies.
4. Plans or participates in menu planning.
5. Manages product costs according to budget requirements.
6. Communicates with service personnel about menu changes and specials.
7. Meets with staff and guests about menu/food production concerns and issues.

Part IV: Job Specification

Education:	High school graduate or equivalent
Skills:	Able to effectively communicate with employees and read/write recipes, calculate costs, and plan menus, employee schedules, and budgets
Experience:	A minimum of three years as a cook or chef; basic knowledge of kitchen equipment and procedures, safety and sanitation regulations, and food production techniques
Physical:	Must be able to work in a fast-paced environment for up to eight hours and be able to lift at least fifty pounds

Writing job descriptions takes care and sensitivity. One challenge is to write them in compliance with the Americans with Disabilities Act (ADA). Always focus on the results or outcomes of a responsibility, identify the essential functions of the position, and describe them in clear and neutral language. **Essential functions** are the key duties that an employee must be able to perform to do the work.

For example, the job description for a sauté cook might include several essential functions:

- Identifies and prepares selected cuts of meat, poultry, shellfish, fish, or vegetables for sautéing
- Sautés and prepares appropriate sauces
- Plates sautéed items with appropriate garnishes
- Maintains a well-organized and sanitary work station

Honoring the spirit and intent of the ADA legislation requires managers to clarify what is involved in specific duties and whether there are alternate ways, called **reasonable accommodations**, to accomplish them. The Equal Employment Opportunity Commission (EEOC) defines reasonable accommodation as a change in the job application process, in the way a job is performed, or to other parts of the job such as employer-sponsored training and benefits that enables a person with a disability to have equal employment opportunities. However, an employer does not have to accommodate a person if doing so would cause an undue hardship. An undue hardship is considered to be an action that is excessively costly, extensive, or substantial in relation to an employer's size and financial resources, or that would fundamentally alter the nature or operation of the business.

Clear job descriptions make it easier to address accommodation questions and to identify parts of a job where accommodation is not possible. For example, if a person is not able to hold or use a knife, he or she cannot be a butcher; there is no way to accommodate the lack of that motor skill. However, a bus person who cannot carry heavy trays can make more trips or be provided with a cart to clear tables.

When developing job specifications, ensure that they identify only **bona fide occupational qualifications (BFOQs)**: the realistic range of skills or credentials needed to perform the essential job functions. For example, a manager cannot specify gender as a job specification unless it is necessary to perform the job, such as "female" for a women's restroom attendant.

Manager's Memo

A job description must clearly define the job. Clarity depends on what information is included and what is left out. Too much information makes the job description difficult to use and to keep current.

Employees must know about other important information that is not normally in the job description, including the operation's policies and procedures. Much of this information is addressed in employee handbooks and policy statements or in other documents including performance standards, checklists, and training materials.

Many policies and procedures apply to most or all employees, and putting them in every job description would be repetitive. Safety and customer service responsibilities are examples of critical responsibilities of all employees. This type of information is often covered by tasks that reference general policies:

> "Follows all policies and procedures as stated in the Employee Handbook."
>
> "Cleans and sanitizes work areas, following the operation's food safety practices."

TRAIN EMPLOYEES TO ATTAIN JOB STANDARDS

Training must be **cost-effective**, meaning it must yield time and money savings that are greater than what it costs. The best training is performance-based, and it is planned and delivered in an organized way so employees will learn how to correctly perform all the tasks in the task list. The term *performance-based* means that the training focuses on what trainees must know and be able to do to perform to the expected levels of work quality and quantity for all tasks in their job descriptions.

The need for training to be performance-based creates certain challenges:

- All tasks in a position must be identified.
- The specific knowledge and skills needed to do each task must be determined.
- Training to provide all of the knowledge and skills to perform each task must be developed.
- **Competencies** (standards of knowledge, skills, and abilities required for successful job performance) must be considered.
- An evaluation process is needed to learn if the training is successful.

The best performance-based training is usually delivered at the job site. The optimal delivery format is one-on-one training sessions.

Training Benefits

There are several benefits of effective training:

- **Improved job performance:** Trainees learn how to perform required tasks more effectively, and this improves their job performance.
- **Reduced operating costs:** Improved job performance helps reduce errors and re-work, which reduces costs.
- **More satisfied customers:** Training can help make employees more customer-oriented.
- **Reduced work stress:** Employees who work correctly will feel better about doing it, and this reduces stress.
- **Increased advancement opportunities:** Competent employees are more likely to receive promotions.
- **Fewer operating problems:** Busy managers can focus on priorities and spend less time on routine problems.
- **Higher work quality:** Effective training identifies quality standards and helps employees work at levels that meet quality requirements.

Manager's Memo

Employees will learn more when they have been prepared for their training:

- Tell them what to expect.
- Explain how the training will help them.
- Allow time for the training.
- Address their concerns.
- Indicate that the training will directly relate to their work.
- Tell how the trainees will be evaluated.
- Express confidence that they will be successful.

On-the-Job Training

On-the-job training is commonly used in the restaurant and foodservice industry. It occurs when a manager or other trainer teaches job skills and knowledge to one trainee, usually at the work site. It is an excellent training method when organized training activities have been developed:

- It uses basic adult learning principles such as considering the trainee's attention span and providing immediate feedback about performance.
- It addresses what the trainee must know and be able to do.
- It can be used to train both new and experienced staff.
- It is well-accepted in the industry.

There are four steps to effective on-the-job training. They are previewed in *Exhibit 7.5*.

TRAINING PREPARATION

Trainers should do several things as they prepare:

- Preview the training objectives. Identify what the trainee will learn.
- Use the task breakdowns. They indicate the training content. It may be helpful to duplicate a copy for the trainee before the training session.
- Consider the training schedule. Know the length of the training activity and where in the overall training plan specific activities should occur.
- Select the training location. When possible, the training should occur in the work area where the task will be performed.
- Assemble training materials and equipment. Get everything needed for the training before it begins.
- Prepare the trainee. A new employee should know that the training will help him or her perform all job tasks correctly. An experienced staff member should understand that the training will provide the knowledge and skills needed to perform a task differently or to learn a new task.
- Determine what the trainee already knows. If a piece of equipment must be operated, ask the trainee to demonstrate its proper operation. If the trainee can do so, this part of the training is not necessary. If the trainee cannot operate the equipment, training should address equipment operation.

TRAINING PRESENTATION

The **training lesson** consists of the information and methods used to present one training session. In addition, the trainer will find that the task breakdown completed during position analysis will be useful.

THINK ABOUT IT . . .

Some managers think they use good on-the-job training tactics when they ask a new employee to follow or "shadow" another employee who has worked in the job for a while. However, this unplanned type of training can seem disorganized to the new employee. For example, the trainee may begin to learn one task, and then must move on to another task before they have truly mastered the first.

How might it feel to be trained by someone who didn't know how to train and made it clear that they thought training you was a bother and an interruption?

Exhibit 7.5

ON-THE-JOB TRAINING PROCESS

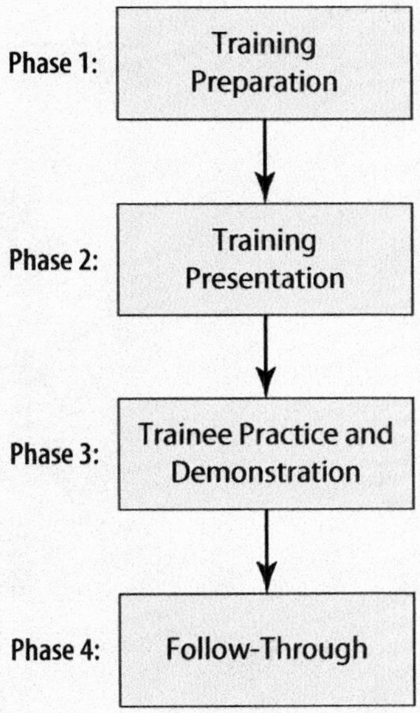

To review effective training presentation practices, think about how a manager might train an employee to conduct a physical inventory in the storeroom. The trainee might be given a copy of the job description that confirms that taking inventory is part of the trainee's job. Its importance might be justified: "A physical inventory count is done every month because it helps us figure out how much we spent for food during the month."

The trainer might then review a copy of the job breakdown for the task: "To do a physical inventory, one person counts the number of cases, cartons, or bags of each product. A second person confirms the inventory quantity and enters it onto a worksheet. This process is continued until all products are counted, and this usually takes about one hour."

After this explanation, the trainer shows the trainee how the storage area is organized and reviews how the inventory worksheet is completed. The counting process is reviewed, along with strategies to ensure products are correctly shelved to promote product rotation.

The trainer then reviews the first step, answers questions, and allows the trainee to repeat, practice, or demonstrate the step. If necessary, the steps are repeated until the trainee learns them correctly.

As the presentation evolves, trainers follow several principles:

- Speak in simple terms.
- Present simple tasks before more complex activities.
- Explain and demonstrate tasks slowly and clearly.
- Use a questioning process, asking open-ended questions such as "Why do you think we count full cases of each item before opened cases?"
- Emphasize the task breakdown during the training.
- Provide clear and well-designed instructions for each task.

Exhibit 7.6

TRAINEE PRACTICE AND DEMONSTRATION

Several principles are important as the trainee practices and demonstrates learning:

- Ask the trainee to repeat or explain key points.
- Have the trainee demonstrate or practice the task. If practical, the trainee should practice each step enough times to learn the "basics" before moving on to the next step. Also, if possible, the steps should be taught and practiced or demonstrated (*Exhibit 7.6*) in the order necessary to perform the task.

- When a task is difficult to learn, progress may be slow. Then the trainee may need to repeat each step more often to build speed and to do it more consistently and correctly. Some trainees learn faster than others. This principle is easy to incorporate into on-the-job training because individual training can be presented at the speed judged best for the trainee.
- Acknowledge correct performance before addressing performance problems.
- Praise trainees for proper performance with at least a compliment or thank-you.

FOLLOW-THROUGH

The final step in on-the-job training is for the trainer to ensure that the training has been effective. For example, at the end of the session, the trainee should be asked to perform each step in the task in sequence. The trainer should encourage him or her and ask questions. It is also useful for the trainer to provide reinforcement about a trainee's positive attitude and improved skills and knowledge.

Close supervision immediately after training and occasional supervision after a task has been mastered can help ensure that the trainee is able to consistently perform the task correctly. Trainers should request that trainees always perform the task in the way they have been trained, and trainees should be asked to keep copies of the training materials provided for later use if needed.

Training evaluation is useful for determining the extent to which the training was successful. Evaluation can also help identify the strengths and weaknesses of training and collect information to help justify future programs.

There are several possible training evaluation methods:

- **Objective tests:** These can be written, oral, or skill-based. A separate test should be used for each training session.
- **Observation of after-training performance:** Managers, supervisors, and trainers can "manage by walking around" and note whether the knowledge and skills taught are being applied.
- **Interviews with trainees and trainers:** The use of open-ended questions such as "What do we do with the leftovers from our buffet?" may provide useful input to evaluate training.
- **Exit interviews:** Formal and even informal conversations with staff members leaving the operation can provide information helpful for training evaluation.

Analysis of customer comments and use of mystery shoppers can also help evaluate training programs.

Manager's Memo

If an employee attends a training session and misses only 2 of 20 questions on an after-training test, most managers would probably assume the training was successful. The trainee answered 90 percent of the questions correctly (18 correct questions ÷ 20 questions = 90 percent correct). However, the training could have been a waste of money and time if the trainee already knew the concepts covered by the 18 questions answered correctly.

Trainers who use pretests and posttests address this concern. They identify key training concepts and give a before-training test (pretest) to learn a trainee's existing knowledge or skills. They then use the same test at the end of the training (post-test). Improved scores measure training effectiveness. Trainers who also assess training several months after the training is completed can learn whether training content was retained to be used on the job.

Documentation is a final part of training. Training information should be maintained in each trainee's employee file:

- Name of trainee
- Training dates
- Training topics
- Notes about successful completion
- Other information as applicable

USE CHECKLISTS

Checklists can be used to help ensure that standards are met and routine procedures are consistently performed throughout different shifts. Many types of checklists are used for many different purposes. For example, some are designed to remind employees to perform all necessary tasks in a timely manner. Others are used to ensure quality standards are maintained and for record-keeping purposes.

Job standards should be developed for all positions. For example, task lists identify the specific tasks that an employee working in a position must be able to do. The job breakdowns for each task tell how each task should be done. They form the foundation for the position's job standards. The establishment's staff plan, menus, National Restaurant Association position guides, organizational chart, and even the mission statement may also help determine the job standards for each position.

Exhibit 7.7

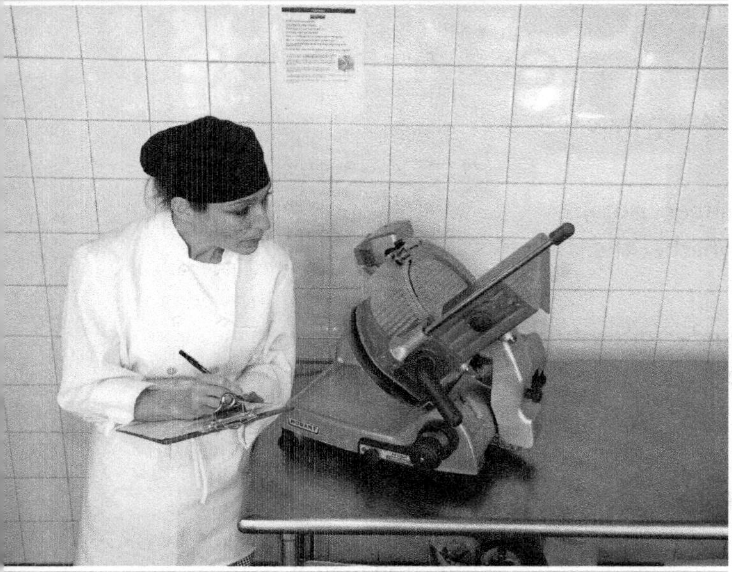

The use of well-planned checklists helps confirm that the tasks that are supposed to be done are being done. Checking on all parts of the operation helps ensure a consistent approach to taking care of customers and the facilities and should become a regular part of daily work.

Opening and Pre-Shift Checklists

Opening and pre-shift checklists are used to ensure that everything has been prepared so the operation will be ready to serve customers when the shift begins. Managers and employees may use different checklists, as shown in *Exhibit 7.7.* However, one or more managers should do a "walk-through" of every area. This will confirm that all activities on the checklists have been satisfactorily completed.

Each manager must ensure that each scheduled employee has arrived at work on time, is prepared to work, and is dressed appropriately. Greeting employees as they enter the establishment helps managers track who has come in and sets a good tone for a hospitality business.

The kitchen manager or head chef must be certain food-production staff are ready to prepare and serve food. An opening or pre-shift checklist for the back of the house commonly addresses several concerns:

- All equipment is clean, sanitized, and ready for use.
- Refrigeration equipment is working properly.
- Equipment such as ovens and deep fryers are ready to be turned on when needed.
- Ingredients required for the first recipes to be produced are removed from storage and transported to work stations.
- All *mise en place* (a French term for having everything ready for preparation) has been done.
- Menu specials are entered into the point-of-sale (POS) system.
- The production sheet that specifies the food items to be produced for the shift has been prepared and posted or explained.

THINK ABOUT IT . . .

Checklists can be helpful for managing a shift. However, managers must still supervise work, communicate with employees, monitor events, and make plans.

What management activities might be difficult to monitor with a checklist?

Mid-Shift Checklists

During a shift, the manager should review the operation of all work areas to ensure staff are doing their work in an efficient manner and that job standards are being attained. In addition, he or she confirms that the physical facilities are being maintained to food safety standards without any "shortcuts" being taken. Sometimes, when production needs are greater or less than estimated, a manager or the chef adjusts staffing by calling in a worker or releasing someone from work.

Managers typically have many things to keep track of, and many use a mid-shift checklist to remind them about critical activities. However, they know that checklists are only reminders and that following a checklist during a shift is no substitute for ongoing monitoring.

In the back of the house, the chef or another manager monitors food preparation, food safety, staffing, and quality standards, among many other concerns. The chef or manager should continuously evaluate the speed with which food is prepared, the extent to which people are working effectively, the level of teamwork, the preparation of food, and the operation of equipment. Ultimately, the chef or manager is responsible for ensuring that quality and food safety standards are met. Food safety is a significant aspect of this managerial role, and the manager has numerous employee-related concerns:

- Work stations are clean and safe from cross-contamination.
- Refrigerators and freezers are monitored to ensure that they are maintaining the proper temperatures.
- HACCP logs are completed as required.

Other common mid-shift tasks for the back of the house relate to production. For example, the production line has the required ingredient par levels or the ingredients are in easy-to-reach locations so food can continue to be prepared quickly and efficiently. Managers must also reorder food products as needed and update production sheets.

Shift-End and Closing Checklists

When a shift is ending or an operation is closing for the day, many activities become important to ensure that all necessary record keeping is completed and that the establishment will be ready for the next shift. Many of the same areas that were checked before the shift will need to be checked again. With well-designed checklists and reliable employees, these tasks happen almost automatically. However, it is still important for the manager to verify all activities were completed to avoid any problems.

Kitchen employees should have properly cleaned all work areas and the equipment in them. Small equipment and tools must be cleaned, properly sanitized, and made ready for the next shift. The manager must ensure that the quantities of items produced correspond to the number of items sold, taking leftovers into account. Please note that if the next shift begins immediately, completing all these tasks may not be necessary or even possible depending on operating procedures and business volume.

A shift-end or closing checklist for the production area might address numerous concerns:

- Production sheets are completed and compared to the actual quantity sold to discover any potential problems.
- All food is wrapped, labeled, and dated.
- All appropriate equipment such as ovens, ranges, hoods, and dishwashers are turned off.
- All food is properly cooled and stored.
- Refrigerators and freezers are secured.
- Closing checklists for each station in the kitchen are completed.
- The floor is clean.
- The trash is removed.

A shift-end or closing checklist for the front of the house is also important and will include common activities. Examples involve completion of all side work and cleaning and restocking side stands. Soda guns, coffee machines, and other beverage dispensers must be broken down and cleaned. Tables may be reset for the next meal, the floor must be cleaned, and the lighting must be adjusted or turned off. Trash containers must be emptied, dirty linens must be removed to the designated location, menu boards may need to be cleared, and self-serve stations must be cleaned and stocked.

EVALUATE EMPLOYEES AGAINST JOB STANDARDS

The best managers have done extensive planning to develop job standards and to ensure that their employees are trained and have the equipment and tools necessary to attain them.

When employees do not meet job standards, it may be due to inadequate training, tools, or equipment. The problem may also be caused by a poorly written job standard. Perhaps some job activities have changed and the job standards are out-of-date. For example, some receiving tasks may now be computerized and the task breakdown still describes the manual system. Job standards must be changed to keep up with the tasks employees must perform.

Managers must confirm that job standards are met and to take corrective action when they are not. Coaching and problem solving are two activities that managers can use when corrective action is required.

Coaching

Coaching occurs when managers emphasize that employees are working correctly (positive coaching) and when they discourage employees from working incorrectly (negative coaching). Coaching is not usually a formal activity, and it often involves simple conversations with employees as they work, as shown in *Exhibit 7.8*.

Exhibit 7.8

Positive coaching occurs when a manager observes an employee doing something correctly. For example: "Joe, you're really doing a great job with that new recipe. Thanks a lot; I really appreciate it." Negative coaching involves correcting an employee. For example: "Joe, I know you are using a new recipe, and there is one step that we should still work on together." After the step is discussed and demonstrated, the manager can conclude the coaching discussion: "Thanks, Joe; I appreciate your hard work, and I hope this step doesn't cause any more problems."

THINK ABOUT IT . . .

Employees like to know how their supervisor feels about their work, just as most students want to know what their teachers think.

How do you feel when a supervisor or teacher discusses your performance?

The best managers remember the idea of praising employees in public. They also remember to offer criticism in private.

Almost all employees want to know how their supervisor views their work. They like coaching input. Since it costs no money and takes so little time, it should be a "must" strategy used by all managers.

Simple coaching can correct many minor staff performance problems. The use of several principles can help ensure coaching is effective:

- Be tactful. If corrective actions are required, focus on the employee's behavior, not on the staff member himself or herself.
- Emphasize the positive. Managers who interact with well-trained and motivated employees have many opportunities to provide positive rather than negative coaching.
- Demonstrate and review appropriate procedures. Spend more time showing the correct way to do something than complaining about incorrect performance.
- Explain reasons for changes. Explain reasons from the employee's point of view; for example: "Joe, we have changed how this is done. There will be fewer errors and less stress for you."
- Keep communications open. A workplace that supports ongoing coaching conversations reduces concerns such as "What does the boss want to talk to me about now?"
- Provide professional development opportunities. These include education and training provided to staff members to improve their present job skills and knowledge, and to prepare them for other positions. Managers can reinforce desired performance with a suggestion that it can lead to additional training and professional development opportunities.
- Allow employees to contribute to their work. Employees can be asked for their ideas when work procedures are evaluated, revised, and implemented. They can also be asked about suggestions for addressing operating challenges.
- Conduct negative coaching in private. Praise employees in public and conduct conversations to improve performance in private.
- Evaluate the work of employees by comparing their performance against standards in task breakdowns. Do not compare one employee's work to that of another employee.
- Establish and agree on time frames for corrective action. If performance is not acceptable, the manager and employee should agree on what must be done to improve it. Then a schedule for acceptable performance or for a review of performance can be determined.

- Ask employees how their performance can be improved. Providing them with task breakdowns and asking for improvement suggestions can be very useful.

Problem Solving

Managers may need to address problems during work shifts that cannot be resolved by simple coaching. They must do what is best to get through the shift and then use a more formal problem-solving method to take corrective action.

Many problems can be resolved by considering existing policies and procedures. Some problems, however, create special challenges and require creative ways to address them.

Problem solving is a process that uses a logical series of activities to determine a course of action. The method shown in *Exhibit 7.9* provides a practical plan that managers often use.

Exhibit 7.9

PROBLEM-SOLVING MODEL

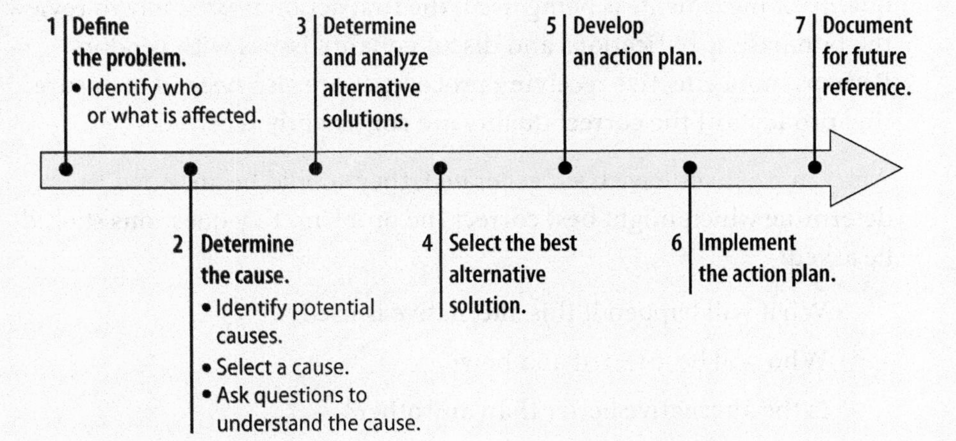

Step 1: Define the Problem

The best way to define a problem is to think about who or what is affected, and a questioning process can be used for this purpose. Depending on the problem, groups of persons including employees, managers, owners, or even customers could be asked to help explain the problem.

Manager's Memo

Most managers monitor their food cost by using a food cost percentage (Food cost ÷ Food revenue = Food cost percentage). If the food cost percentage is too high, managers typically think the problem relates to a food problem. For example, they may consider inventory theft and evaluate product portion sizes.

Managers should consider that a food cost percentage can also be too high because the revenue is too low (see the previous calculation). Therefore, it can be caused by theft of revenue or embezzlement (stealing money from the operation).

This example makes the point that it is critical to properly define a problem before solving it. Then the remaining steps proceed on a solid foundation.

Step 2: Determine the Cause

Problems affecting job standards are often caused by human error or by a failure to follow procedures such as using standardized recipes. Operations are made up of a series of complex systems including those for purchasing, production, customer service, and cash handling. Problems can occur when one or more of these systems breaks down or was not carefully developed in the first place.

One way to identify the cause of a problem is to think about potential causes, and managers often use a questioning process to do this. For example, if there is a food quality problem, is it because purchase specifications are not being followed? Could it be because employees are taking "shortcuts" when using standardized recipes?

Problems may also arise with higher-than-expected food costs, and there are many possible reasons that might be investigated. For example, are standardized recipes being followed or have food prices increased? Is employee food theft causing higher food costs or is there excessive waste or spoilage?

Step 3: Determine and Analyze Alternative Solutions

An alternative is a possible solution to a problem. For example, if the manager thinks a food quality problem has arisen because the wrong quality of ingredients is being used, the first action is probably to review the purchase specifications and discuss quality issues with vendors. Perhaps more effective receiving procedures are also needed to ensure that products of the correct quality are consistently received.

After alternatives have been generated, they should be analyzed to determine which might best correct the problem. Key questions should be asked:

- What will happen if this alternative is used?
- Who will be affected and how?
- Is the alternative better than any other?
- Is it cost-effective? Will the solution cost more than the problem?
- Can the alternative be revised to cost less and still fix the problem?
- Is it reasonable? Does the alternative have a good chance of succeeding?

Step 4: Select the Best Alternative Solution

Use a questioning process to find the best alternative to eliminate or reduce the problem. Which alternative is best? Is there a way to combine parts of two or more alternatives to yield an even better alternative? An ineffective solution will create the need to repeat the problem-solving process, and the negative impact of the problem will also continue.

THINK ABOUT IT . . .

Albert Einstein said "We can't solve problems by using the same kind of thinking we used when we created them."

Why is asking questions important to solving problems? Why could using a team be important?

Step 5: Develop an Action Plan

After the solution has been chosen, an action plan must be developed. An **action plan** is a series of steps that will be taken to resolve the problem. The action plan for a food quality problem might be to ask the purchaser to revise a purchase specification using current information from the product vendor. A due date for revising, testing, and implementing the purchase specification would be an important part of the action plan.

Step 6: Implement the Action Plan

Communicate the action plan and its expected outcomes to everyone involved. This ensures that all parties know what must be done and how they will be impacted.

Step 7: Document for Future Reference

Before beginning the problem-solving activity, managers can review previously filed solutions for similar events that may be helpful in resolving the current situation. Documentation will also be helpful in updating job standards. For example, a revised purchase specification or standardized recipe must be implemented and consistently used to help reduce the failure to meet job standards. Changes in work procedures may affect task breakdowns and even job descriptions.

The purpose of documentation is to record information for future use. Changes in job procedures and activities used to most consistently meet job standards should be incorporated into training, coaching, and ongoing supervisory responsibilities.

CONTROL LABOR COSTS DURING PRODUCTION

Controlling labor costs is an important part of every manager's job. Some inexperienced managers may think that the best way to control labor costs is to schedule fewer labor hours. However, professional managers understand that their customers are best served when the correct number of well-trained employees is matched with the business volume expected.

Factors Affecting Labor Costs

The best labor cost control systems ensure that production and service standards and the operation's financial goals are met. The quality and number of staff required to do that vary based on several factors:

- **The menu:** The menu is probably the most important factor in controlling labor costs. There can be just a few or many menu items that are time-consuming to produce. Managers must provide the menu variety necessary to ensure the operation's popularity and success. At the same time, they must schedule employees who can prepare and serve that menu variety while meeting quality standards.

Exhibit 7.10

- **Food-preparation methods:** Few operations prepare all menu items "from scratch" using unprepared ingredients. Some types of convenience food are used to reduce labor costs (*Exhibit 7.10*), and managers must select convenience food items based on acceptable quality as the first consideration, as well as labor savings. A manager who buys inferior products to save labor costs will quickly learn that customers will not accept poor-quality food items. Spending less for products that are not acceptable does not make business sense, nor does it save money.
- **Quality of training:** Training improves the knowledge and skills of employees and can help increase productivity and lower labor costs. New employees must be well trained to perform to job standards. Then, as they become more productive, they can work more cost-effectively.

Exhibit 7.11

FIVE STEPS TO CONTROLLING LABOR COSTS

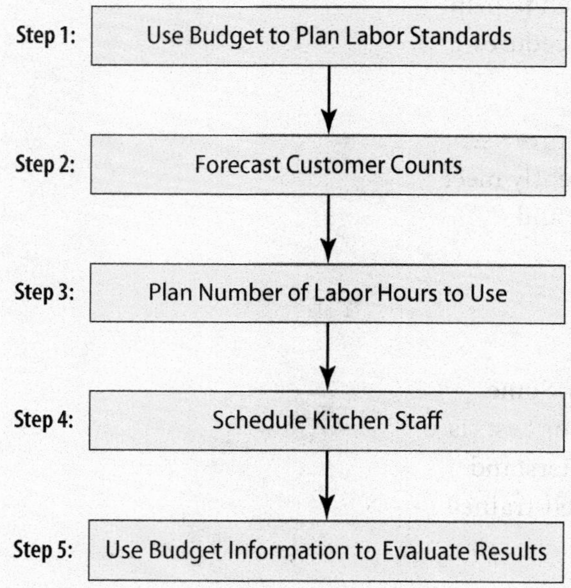

Five Steps to Controlling Labor Costs

Managers can use an organized process to control labor costs. It involves five basic steps previewed in *Exhibit 7.11*.

Step 1: Use Budget to Plan Labor Standards

Most managers use operating budgets that estimate revenues and expenses for a specific period for two purposes. Budgets help them estimate their revenues and expenses, and they can then measure actual results against the budget standards to learn how well their plans worked out.

Budgeted labor cost goals cannot be met unless they are used to help plan employee schedules. Therefore, the approved operating budget serves as a roadmap to help managers spend no more than the budget limits for labor costs.

Consider how Zellie, the manager of Grey's Family Restaurant, uses her operating budget to manage her labor costs. She has a monthly budget of $27,000 for waged employees (those paid by the hour) excluding benefits for the month of April. Her operation is open for dinner six days each week, and she plans the employee schedule on a weekly basis.

First, Zellie determines the number of days for which the employee schedule is needed. Looking at her calendar, Zellie notes that there are four Sundays in April this year. Therefore, there will be 26 workdays (30 days in April − 4 Sundays = 26 workdays).

Zellie can now determine how much she can spend on wages on an average day:

$27,000	÷	26	=	$1,038.00 (rounded)
Wage budget for April		Workdays in April		Average daily allowable wages

If the average hourly wage, not including benefits, is $17, Zellie can schedule no more than 61 labor hours in the kitchen for an average day in April:

$1,038.00	÷	$17.00	=	61 (rounded)
Average daily wages		Average hourly rate		Average allowable daily hours

Her labor budget also suggests that no more than 366 waged hours can be worked in an average week:

61	×	6	=	366
Average allowable daily hours		Days in work week		Average allowable waged hours per week

Step 2: Forecast Customer Counts

Now that Zellie knows how many labor hours can be used each day for waged kitchen employees, she can develop the employee schedule. She must first forecast how many customers she will serve during the week. Procedures for forecasting customer counts were discussed in chapter 5. The process involves using point-of-sale (POS) system information to determine previous sales. This historical information is then adjusted as necessary to make forecasts based on factors such as special community events and holidays.

Step 3: Plan Number of Labor Hours to Use

Employee schedules can be developed by considering the number of labor hours needed to produce meals for the forecasted number of customers. However, a minimum number of labor hours is needed to operate the kitchen. Then additional hours are required as the customer count increases.

Zellie's kitchen opens at 2:00 p.m. to do preparation work for the evening meal that begins at 5:00 p.m. Customers are served from 5:00 p.m. to 10:30 p.m., and then it takes until about midnight to clean up after the kitchen closes. This means that at least 10 waged cook hours will be needed from 2:00 p.m. to midnight, and a dish washer will be needed for *at least* 6.5 hours from 5:00 p.m. to 11:30 p.m. Employees in other back-of-the-house positions may also be needed to open the kitchen. It is only after a minimum threshold of customers is served that additional labor hours are needed.

OPEN FOR BUSINESS

MANAGER'S MATH

Use the following information to answer the questions:

- Clay's Lunch Spot is open seven days a week for breakfast and lunch. It is open 6:00 a.m.–1:00 p.m., so all meals are prepared during a single shift.
- The wage budget for food-production employees (excluding benefits) for September is $37,500, and the average hourly wage is $14.75.

1. How much money can be spent each day on wages for food-production employees?
2. How much can be spent each week for wages for food-production employees?
3. How many food-production hours can be used each day?

Answer 1:

$37,500	÷	30	=	$1,250
September labor budget		*Days*		*Daily wages*

Answer 2:

$1,250	×	7	=	$8,750
Daily wages		*Days in a week*		*Weekly wage spending*

Answer 3:

$1,250	÷	$14.75	=	85 (rounded)
Daily wages		*Avg. hourly rate*		*No. of food-production hours*

Manager's Memo

It is easy to calculate the number of labor hours available to meet labor cost goals. It is another thing to have enough labor hours available when needed. This concern is best considered when the budget is developed. If managers forecast many customers, budget goals for labor costs will be higher. The challenge of producing food with fewer labor hours than needed is minimized when the forecast is accurate.

What happens if the budget estimate is wrong? More labor hours will be needed, and labor costs will be higher than budget goals. Managers must learn from their errors and try to develop more accurate budget estimates. They must also understand that while labor dollars are higher because of greater production volume, revenues will also be higher than expected.

Managers must realize that their goal is not to minimize labor costs. It is to generate planned profits by consistently meeting quality standards.

Recall that Zellie learned in step 1 that she can, on average, schedule 61 hours for waged employees each day. Based on her experience, more hours will be scheduled for busy days, and fewer hours will be needed for low-volume days. However, only 366 hours can be scheduled for the week to remain in line with the month's budgeted labor costs for waged kitchen employees.

61	×	6	=	366
Hours per day		**Days per week**		**Hours**

Step 4: Schedule Kitchen Staff

Managers know that their scheduling skills impact several things:

- The quality of products served to customers
- The employees' level of job satisfaction
- The operation's profitability
- The view of upper management about the manager's abilities

Many managers use software tools to create employee schedules. They can preload employee data including requested days off, vacations, maximum allowable hours, restrictions on when hours can be worked for minors, employee time preferences, and other factors.

Effective employee schedules should contain important information:

- The dates covered by the schedule
- The days of the week covered by the schedule
- Employees' first and last names
- Scheduled days to work
- Scheduled days off
- Scheduled start and stop times (indicate a.m. and p.m.)
- Total hours to be worked
- Requested vacation or personal days off
- Date the schedule was prepared and who prepared it

Exhibit 7.12 shows a spreadsheet detailing the employee work schedule Zellie created for February 2 to 8. Please note that all employees at her operation receive a half-hour unpaid break that is built into their scheduled hours. Zellie's own working times are not listed because she is paid a salary and her actual working hours depend on when she is needed.

Exhibit 7.12

SAMPLE EMPLOYEE SCHEDULE WORKSHEET

Grey's Family Restaurant From: February 2 through 8

Area/ Employee	Monday	Tuesday	Wednesday	Thursday	Friday	Saturday	Sunday	Total Labor Hours	
	2/2	2/3	2/4	2/5	2/6	2/7	2/8	Scheduled	Actual
(1) **Kitchen**	(2)	(3)	(4)	(5)	(6)	(7)	(8)	(9)	(10)
Jerrod	2:00 p.m.– 9:30 p.m.	OFF	2:00 p.m.– 9:30 p.m.	2:00 p.m.– 9:30 p.m.	2:00 p.m.– 9:30 p.m.	2:00 p.m.– 9:30 p.m.	OFF	35.0	
Chelita	5:30 p.m.– 12:00 a.m.	5:30 p.m.– 12:00 a.m.	5:30 p.m.– 12:00 a.m.	OFF	5:30 p.m.– 12:00 a.m.	2:00 p.m.– 8:00 p.m.	OFF	29.5	
Nissor	OFF	2:00 p.m.– 7:30 p.m.	2:00 p.m.– 7:30 p.m.	2:00 p.m.– 7:30 p.m.	2:00 p.m.– 7:30 p.m.	4:00 p.m.– 11:00 p.m.	OFF	26.5	
Sam	5:30 p.m.– 12:00 a.m.	OFF	5:30 p.m.– 12:00 a.m.	5:30 p.m.– 12:00 a.m.	5:30 p.m.– 12:00 a.m.	5:30 p.m.– 11:30 p.m.	OFF	29.5	
Carlo	4:00 p.m.– 10:30 p.m.	4:00 p.m.– 10:30 p.m.	OFF	3:00 p.m.– 10:30 p.m.	3:00 p.m.– 10:30 p.m.	4:00 p.m.– 9:00 p.m.	OFF	30.5	
Lenny	2:00 p.m.– 9:30 p.m.	2:00 p.m.– 9:30 p.m.	2:00 p.m.– 9:30 p.m.	2:00 p.m.– 9:30 p.m.	OFF	OFF	OFF	28.0	
Carmel	OFF	2:00 p.m.– 8:30 p.m.	2:00 p.m.– 8:30 p.m.	2:00 p.m.– 8:30 p.m.	2:00 p.m.– 8:30 p.m.	2:00 p.m.– 8:30 p.m.	OFF	30.0	
Alex	2:00 p.m.– 9:30 p.m.	2:00 p.m.– 9:30 p.m.	OFF	2:00 p.m.– 9:30 p.m.	2:00 p.m.– 9:30 p.m.	2:00 p.m.– 9:30 p.m.	OFF	35.0	
Dishroom									
Larry	OFF	5:00 p.m.– 11:00 p.m.	5:00 p.m.– 11:00 p.m.	5:00 p.m.– 11:00 p.m.	5:00 p.m.– 11:00 p.m.	5:00 p.m.– 11:00 p.m.	OFF	27.5	
Ray	6:00 p.m.– 12:00 a.m.	OFF	6:00 p.m.– 12:00 a.m.	6:00 p.m.– 12:00 a.m.	6:00 p.m.– 12:00 a.m.	6:00 p.m.– 12:00 a.m.	OFF	27.5	
Art	6:00 p.m.– 11:30 p.m.	6:00 p.m.– 11:30 p.m.	6:00 p.m.– 11:30 p.m.	OFF	4:00 p.m.– 11:30 p.m.	4:00 p.m.– 11:30 p.m.	OFF	29.0	
Shayne	OFF	1:00 p.m.– 5:00 p.m.	4:30 p.m.– 9:00 p.m.	1:00 p.m.– 11:00 p.m.	VACATION	VACATION	OFF	17.0	
Kodi	6:00 p.m.– 11:30 p.m.	5:30 p.m.– 11:30 p.m.	OFF	5:30 p.m.– 11:00 p.m.	6:30 p.m.– 11:30 p.m.	OFF	OFF	20.0	

Prepared: 1/26 Approved by: RR 1/27 **Total Hours 365**

Exhibit 7.13

The employee schedule shown in *Exhibit 7.12* shows the days employees will work and their shifts' start and end times. Column 9 indicates the total hours each employee is scheduled to work during the week. The total at the bottom of the column is 365 hours, which is 1 hour less than the maximum number of hours that Zellie can schedule based on her labor budget. Column 10 will be used to report the number of hours each employee actually worked.

Sometimes, as a result of unexpected business volume, unexpected employee absences, or employee separations, schedule changes are needed. If so, affected employees should be quickly informed about the changes. In some operations, employees who begin work at the scheduled time and then are asked to check out early, or who are not needed at all, are paid a minimum amount regardless of the time they actually worked. Know the company's policies before making schedule adjustments and communicate schedule changes appropriately.

In many operations, schedules are posted in one or more central areas (*Exhibit 7.13*), and they are sometimes included with paychecks. Increasingly, managers email schedules or make them available on the establishment's intranet system.

Managers should monitor schedules carefully and adjust them when certain events arise, such as the following:

- Significant changes in business volumes
- Unexpected employee separations
- Employee call-ins or no-shows
- Changes in operating hours or work assignments

Step 5: Use Budget Information to Evaluate Results

After managers have developed, distributed, and implemented the employee schedule, the employees will work their assigned shifts, and managers can evaluate the results. This section reviews procedures to evaluate planned (scheduled) labor hours against the labor cost standard estimated in the approved budget.

If the budget goal is not met, the amount by which actual performance differs from the goal is called a budget **variance**: the difference between an amount of revenue or expense indicated in the budget and the actual amount of the revenue or expense. *Exhibit 7.14* shows Zellie's employee schedule worksheet with column 10 completed. This column indicates the actual hours worked for each waged employee.

OPEN FOR BUSINESS

RESTAURANT TECHNOLOGY

Technology can make employee scheduling for restaurant or foodservice operations of all sizes easier, quicker, and more accurate. For example, online systems allow managers to send schedules to employees by email or text message. Employees can use the system to request time off or trade shifts, and this information will be available on the schedule creation page when the schedule planner wants to use it.

These systems monitor the number of hours employees have actually worked. This is helpful because managers can then discover who is below overtime hours if additional hours must be scheduled because of unexpected business volume. From a legal view, **overtime** refers to hours for which a wage premium of 1.5 times the hourly rate must be paid.

Electronic scheduling systems can track scheduled and actual hours and individual employee attendance trends. All of this information is available to the manager and employees 24 hours a day.

To learn more details about numerous employee-scheduling systems, just type "employee scheduling software" into your favorite search engine.

Exhibit 7.14

SAMPLE EMPLOYEE SCHEDULE WORKSHEET SHOWING ACTUAL HOURS WORKED

Grey's Family Restaurant From: February 2 through 8

Area/ Employee	Monday	Tuesday	Wednesday	Thursday	Friday	Saturday	Sunday	Total Labor Hours	
	2/2	2/3	2/4	2/5	2/6	2/7	2/8	Scheduled	Actual
(1) **Kitchen**	(2)	(3)	(4)	(5)	(6)	(7)	(8)	(9)	(10)
Jerrod	2:00 p.m.– 9:30 p.m.	OFF	2:00 p.m.– 9:30 p.m.	2:00 p.m.– 9:30 p.m.	2:00 p.m.– 9:30 p.m.	2:00 p.m.– 9:30 p.m.	OFF	35.0	38.0
Chelita	5:30 p.m.– 12:00 a.m.	5:30 p.m.– 12:00 a.m.	5:30 p.m.– 12:00 a.m.	OFF	5:30 p.m.– 12:00 a.m.	2:00 p.m.– 8:00 p.m.	OFF	29.5	29.5
Nissor	OFF	2:00 p.m.– 7:30 p.m.	2:00 p.m.– 7:30 p.m.	2:00 p.m.– 7:30 p.m.	2:00 p.m.– 7:30 p.m.	4:00 p.m.– 11:00 p.m.	OFF	26.5	29.5
Sam	5:30 p.m.– 12:00 a.m.	OFF	5:30 p.m.– 12:00 a.m.	5:30 p.m.– 12:00 a.m.	5:30 p.m.– 12:00 a.m.	5:30 p.m.– 11:30 p.m.	OFF	29.5	32.0
Carlo	4:00 p.m.– 10:30 p.m.	4:00 p.m.– 10:30 p.m.	OFF	3:00 p.m.– 10:30 p.m.	3:00 p.m.– 10:30 p.m.	4:00 p.m.– 9:00 p.m.	OFF	30.5	30.5
Lenny	2:00 p.m.– 9:30 p.m.	2:00 p.m.– 9:30 p.m.	2:00 p.m.– 9:30 p.m.	2:00 p.m.– 9:30 p.m.	OFF	OFF	OFF	28.0	28.0
Carmel	OFF	2:00 p.m.– 8:30 p.m.	2:00 p.m.– 8:30 p.m.	2:00 p.m.– 8:30 p.m.	2:00 p.m.– 8:30 p.m.	2:00 p.m.– 8:30 p.m.	OFF	30.0	31.0
Alex	2:00 p.m.– 9:30 p.m.	2:00 p.m.– 9:30 p.m.	OFF	2:00 p.m.– 9:30 p.m.	2:00 p.m.– 9:30 p.m.	2:00 p.m.– 9:30 p.m.	OFF	35.0	37.5
Dishroom									
Larry	OFF	5:00 p.m.– 11:00 p.m.	5:00 p.m.– 11:00 p.m.	5:00 p.m.– 11:00 p.m.	5:00 p.m.– 11:00 p.m.	5:00 p.m.– 11:00 p.m.	OFF	27.5	29.0
Ray	6:00 p.m.– 12:00 a.m.	OFF	6:00 p.m.– 12:00 a.m.	6:00 p.m.– 12:00 a.m.	6:00 p.m.– 12:00 a.m.	6:00 p.m.– 12:00 a.m.	OFF	27.5	34.5
Art	6:00 p.m.– 11:30 p.m.	6:00 p.m.– 11:30 p.m.	6:00 p.m.– 11:30 p.m.	OFF	4:00 p.m.– 11:30 p.m.	4:00 p.m.– 11:30 p.m.	OFF	29.0	32.0
Shayne	OFF	1:00 p.m.– 5:00 p.m.	4:30 p.m.– 9:00 p.m.	1:00 p.m.– 11:00 p.m.	VACATION	VACATION	OFF	17.0	19.0
Kodi	6:00 p.m.– 11:30 p.m.	5:30 p.m.– 11:30 p.m.	OFF	5:30 p.m.– 11:00 p.m.	6:30 p.m.– 11:30 p.m.	OFF	OFF	20.0	20.0
Prepared: 1/26	**Approved by: RR 1/27**						**Total Hours**	**365**	**390.5**

In *Exhibit 7.14*, notice that the scheduled (column 9) and actual (column 10) number of hours worked by each kitchen and dish washing employee is listed. It also shows at the bottom of columns 9 and 10 that while 365 labor hours were scheduled, 390.5 hours were actually worked. This created a variance of 25.5 hours:

390.5	–	**365**	=	**25.5**
Actual hours		**Scheduled hours**		**Hours variance**

There may be an acceptable reason for the variance such as revenue being higher than budgeted and additional labor hours being required to handle the extra business. If this is not the case, the managers should use the problem solving approach discussed in this chapter. Then they can determine the reasons for the higher-than-expected usage of labor hours.

Is this variance significant? A quick calculation will answer this question. Remember that the average hourly wage for each staff member scheduled was $17 without benefits.

The 25.5 unscheduled hours that were actually worked cost $433.50:

25.5	×	**17**	=	**$433.50**
Hours variance		**Average hourly wage**		**Additional labor costs per week**

If this does not seem significant, consider that if these work hours continued for the entire year, there would be a wage variance of approximately $22,542:

52	×	**$433.50**	=	**$22,542**
Weeks per year		**Additional labor costs per week**		**Additional labor costs per year**

Also, because some benefits are tied directly to labor costs, they would also increase. Zellie must investigate this variance for waged labor hours. She now knows the amount of excess labor hours, and this is the starting point for discovering and resolving problems.

SUMMARY

1. **List and describe the procedures for performing a position analysis to create job standards.**

   A position analysis identifies each task that is part of a position and explains how it should be done. Four steps are involved: develop a task list, plan task breakdowns, consider performance standards to explain what will happen when the task is done correctly, and write a job description.

   Supervisors and experienced workers should be consulted when developing task lists and job breakdowns. Any available written information and observation will also be helpful.

2. **Explain strategies for developing job descriptions.**

   Job descriptions summarize the major tasks in a position. They are used for recruiting and screening, selection and orientation, professional development and training, and performance appraisals. They are also useful for salary administration, safety and security, legal, and union relations purposes.

   Job descriptions should be specific but not so detailed that they will be difficult to maintain if minor job tasks change. They must be in compliance with the Americans with Disabilities Act (ADA) and focus on essential functions an employee must be able to perform.

3. **Describe procedures for implementing performance-based training programs that help employees consistently attain job standards.**

   Performance-based training requires that all position tasks are identified, specific knowledge and skills required for each are known, and training addresses this knowledge and skill base. Competencies are also considered, and an evaluation process is used to measure success.

   Benefits of effective training include improved job performance, reduced operating costs, and more satisfied customers. On-the-job training involves a manager or another trainer teaching job skills and knowledge to trainees at the work site.

   There are four steps to training: preparation, presentation, practice and demonstration, and follow-through. Training evaluation methods include objective tests, observation of after-training performance, interviews with trainees and trainers, and exit interviews.

4. **Explain how checklists can help managers ensure that job standards are being met.**

   Checklists remind employees about tasks that must be performed and help ensure the maintenance of quality standards. They are also used to keep records of tasks performed.

   Opening or pre-shift checklists help ensure employees are ready to serve customers when the shift begins. These checklists address equipment, ingredients, *mise en place,* POS system updates, and food-production requirements.

   Mid-shift checklists help confirm job standards are met as work evolves. Food safety is a significant concern. Checklists should confirm clean and safe work stations, refrigerator and freezer temperatures, and HACCP logs.

   Shift-end and closing checklists address all activities required to end the shift or close the operation. Cleaning and sanitation activities are important, as are tasks to secure the establishment.

5. **Describe procedures for evaluating employees' performance as needed to maintain job standards.**

   Managers use effective coaching when they encourage employees working correctly (positive coaching) or discourage employees from working incorrectly (negative coaching). Coaching is typically an informal activity involving conversations with employees as they work.

Successful coaching involves being tactful, emphasizing the positive, conducting negative coaching in private, demonstrating and reviewing correct procedures, and explaining reasons for changes. Effective managers ask employees about ideas for procedure improvements and evaluate each employee's work against performance standards rather than other employees. They also interact with employees to establish time frames for corrective action.

More significant challenges may require a problem-solving approach. The basic steps are to define the problem, determine its cause, analyze alternative solutions, and select the best alternative to develop and implement an action plan. Documenting the process for future reference is also important.

6. **Describe a five-step process for controlling labor costs during production.**

Labor costs are affected by several factors including the menu, food preparation methods, and quality of employee training.

A five-step process can be used to control labor costs. Step 1 involves using budget information to plan the amount that can be spent for a specific day or shift. Step 2 involves forecasting customer counts. Step 3 requires planning the number of labor hours based on the forecast. Step 4 involves scheduling employees so allowable labor hours are not exceeded; schedules may be adjusted due to certain occurrences. Step 5 is determining whether actual labor hours exceeded budget standards, causing a variance.

APPLICATION EXERCISE

Break into teams of two or three students and brainstorm possible causes and alternatives to resolve the following problems. If time permits, share your causes and alternatives with your classmates.

Problem	Cause of Problem	Alternative Solutions
1. Many customers are complaining that their steaks are under- or overcooked.		
2. Customers are complaining about the portion size of several casserole entrées.		
3. One of the cooks suffers a serious cut during the busiest part of a shift and is taken to a nearby emergency clinic.		
4. Refrigerator temperature is too low when employees begin work on a Sunday morning, when refrigeration repair services are closed.		
5. During a busy shift, a sharp knife is found in the bottom of a wash sink full of water. No injury occurs.		

REVIEW YOUR LEARNING

Select the best answer for each question.

1. **All specific duties involved in performing a job are included in a**
   A. task list.
   B. task breakdown.
   C. performance standard.
   D. job specification.

2. **The major duties of a position are summarized in a**
   A. task list.
   B. job specification.
   C. job description.
   D. performance standard.

3. **Essential functions defined for ADA purposes should be noted in which document?**
   A. Task list
   B. Task breakdown
   C. Job description
   D. Job specification

4. **Which document is useful when on-the-job training is presented?**
   A. Job specification
   B. Task breakdown
   C. Mission statement
   D. Evaluation form

5. **What do pretests and posttests help managers identify?**
   A. Which employees need follow-up observation
   B. What was learned as a direct result of training
   C. Whether trainees can apply training to the job
   D. How trainees evaluate the training received

6. **Which operational checklist most likely addresses *mise en place?***
   A. Opening checklist
   B. Mid-shift checklist
   C. Shift-end checklist
   D. Closing checklist

7. **Which procedure is most useful for effective coaching?**
   A. Emphasize the positive
   B. Discipline employees in public
   C. Compare one employee to another
   D. State how performance affects profits

8. **What is the term for a series of steps taken to resolve a problem?**
   A. Action plan
   B. Follow-through
   C. Task breakdown
   D. Operational procedure

9. **Which is the first consideration when determining whether types of convenience food should be used?**
   A. Use the maximum number of convenience food items.
   B. Eliminate menu items that are labor-intensive.
   C. Schedule the minimum number of labor hours.
   D. Use the operating budget in scheduling labor.

10. **A manager can spend no more than $950 per day for waged kitchen employees. The average hourly rate is $14.75. Approximately how many waged hours can be scheduled per day?**
    A. 34
    B. 44
    C. 54
    D. 64

FIELD PROJECT

Imagine that you are opening a new Italian themed establishment. You will need cooks, bartenders, and food servers. Develop a list of important tasks that employees in each position should know how to do. Hint: *Enter "job description for cooks" (or bartenders or servers) in your favorite search engine.*

Cooks' major job tasks include:

Bartenders' major job tasks include:

Servers' major job tasks include:

8 Communicating with Customers

INSIDE THIS CHAPTER

- Customer Concerns about Nutrition
- Truth-in-Menu Concerns
- Allergies and the Menu
- Responsible Service of Alcoholic Beverages

CHAPTER LEARNING OBJECTIVES

After completing this chapter, you should be able to:

- Explain how effective managers plan ways to incorporate a nutrition emphasis into their menus and food-preparation procedures.
- Review examples of truth-in-menu concerns that must be addressed when menu descriptions are written.
- Describe elements in an organized system for ensuring the health of customers with food allergies.
- Explain concerns about the service of alcoholic beverages.

KEY TERMS

anaphylaxis, p. 231

blood alcohol content (BAC), p. 240

cross-contact, p. 233

food allergy, p. 231

nutrition, p. 226

proof, p. 242

punitive damages, p. 235

truth-in-menu laws, p. 230

CASE STUDY

"This always happens at the busiest times!" John said. The chef at Friendly Bluffs Restaurant was talking to Natasha, the manager. She had just informed him that a customer with a food allergy wanted information about several menu items.

"You know," said John, "it's difficult enough to put several items on the menu that will be OK for customers with food allergies. But now, in the middle of a busy rush, we are supposed to make a special order with clean utensils, supplies, and even work surfaces."

"A customer wants us to prepare an item from scratch while the restaurant is really jammed! There are 125 customers in the place, and he is only one with an issue. We just can't do it."

1. What would you say to Chef John if you were Natasha?
2. What procedures would you have in place to handle orders involving food allergies when the establishment is very busy?

CUSTOMER CONCERNS ABOUT NUTRITION

Managers in almost all segments of the hospitality industry are recognizing increased customer concerns about nutrition. **Nutrition** is the science of food and how it affects the health and well-being of the person who consumes it. Healthy lifestyles that emphasize diet and exercise are increasingly important. Managers understand the need to consider operational changes that address this trend.

Nutrition Is a Trend

The degree of emphasis on nutrition varies by type of customer. For example, some operations target nutrition-conscious diners as their primary market. All or almost all of their menu offerings are selected and prepared with a nutrition focus. Examples include healthcare facilities and short-term residential centers where clients learn how to address personal wellness concerns.

Still other operations cater to a market with significantly more concerns about nutrition. However, they also attract customers who enjoy other food. Examples include spas and athletic clubs.

Perhaps most operations are in a third category where customers vary. Some customers are very concerned about nutrition, while others are not concerned at all. Still others have concerns in between these extremes.

Exhibit 8.1

The extent to which nutrition concerns drive menu planning and recipe development differs in these basic categories of operations. For example, all healthcare facilities and many other noncommercial operations, such as the military and correctional institutions, employ dietitians. Nutrition concerns are high priority. Their food-production staff must be trained to recognize the importance of nutrition and consistently deliver menu items meeting nutrition goals (*Exhibit 8.1*).

Managers in some other operations may work with production staff to develop or modify recipes with a nutrition focus. In some instances, they may purchase the recipes from commercial sources or retain a consultant for this purpose.

The majority of managers and employees in commercial operations are not nutrition experts. They deliver menu items to customers with special nutrition concerns by, for example, modifying preparation and service procedures for existing menu items. Examples of revisions in food-preparation methods were presented in chapter 5.

Planning for Nutrition Concerns

Managers of commercial operations work closely with their food-production staff as menus are planned. The focus of the menu planning team must be on answering the question "What do our customers want at a selling price that promotes value and allows our operation to meet its financial goals?" Increasingly, the answer to that question includes items for nutrition-conscious customers.

There is justification for this response because studies undertaken during the last 10 years have shown that consumers are increasingly interested in where their food comes from and how it is produced. Also, more than 1,600 chefs ranked local produce as the top food trend in a 2009 survey by the National Restaurant Association.

Concerns about nutrition are a very important aspect of this food trend. There are numerous ways to meet these customers' needs, and they are discussed throughout this discussion of nutrition.

MENU PLANNING

It is easier to consider practical ways to alter menu items for nutritional concerns when they are being planned, rather than after the menu is completed. For example, menu planners might decide to offer a "Special Fresh Fish of the Day." In addition to offering different fish varieties, they may also prepare the daily special according to their customers' preference: baked, broiled, pan-fried, deep-fried, or steamed in a special lemon (not butter) sauce.

There are many simple ways to offer menu items of interest to customers with diet concerns. There are three things that managers can do to ensure these alternatives are practical:

- Discuss possibilities with production staff while the menu is being planned. Do not wait until a customer asks "Can you bake this item?"
- Review information in trade publications, attend educational programs sponsored by professional associations, and study the menus of other establishments. Doing this allows managers to keep up with new ideas useful for their operation.
- Consider the trend in nutrition concerns as an opportunity to improve, not a problem to be resolved. Managers who role-model a "can do" attitude are much more effective than those who exhibit a "must do" philosophy. When food-production and service teams work together to meet the needs of this ever-increasing customer market, the operation will be strengthened.

SIMPLE CHANGES

Some simple service changes do not involve recipes, and they are relatively easy to make. For example, salad dressings can be offered on the side rather than

Exhibit 8.2

ladled over the salad. Baked potatoes can be served without butter and sour cream (*Exhibit 8.2*). Some recipes may not require changes if they are prepared on a by-order basis, such as an omelet made with egg whites instead of egg yolks.

As discussed in chapter 5, creative managers and their food-production staff can offer a variety of menu items to nutrition-conscious customers without making extensive modifications in recipes or food production. Other recipes will need more significant modification, such as using applesauce to replace fats or oils in a baking recipe.

NUTRITIONAL CONTENT OF RECIPES

Some managers in healthcare and other facilities need to know the nutritional content of the recipes they use. This information is important when they plan menus or offer menu items for customers on low-sodium (salt) or high-protein diets. Managers in commercial operations may want to learn the nutrition content of some of their recipes to provide nutrition information on the menu. Also, there are local menu labeling regulations in some areas. These regulations provide another reason why it is very important for chefs and even managers to be familiar with the nutritional content of items produced by their recipes.

A six-step procedure can be used by managers who want to determine approximate nutritional content. First, they must determine the standardized recipes for which this information is desired. Some managers may want this information for all recipes, but many want detailed nutritional data for only a few. Managers and food-production staff should select the recipes and confirm that they are current, accurate, and consistently used.

A second step is to design a nutrition worksheet that considers all of the information the manager desires. An example is shown in *Exhibit 8.3*.

Exhibit 8.3

SAMPLE NUTRITION WORKSHEET FOR STANDARDIZED RECIPES

Recipe Name	Recipe No.	Recipe Yield (No. of Servings)	Ingredient Analyzed	Calories	Cholesterol	Total Fat	Sodium
			Total				

Manager's Memo

Many managers in commercial operations have the best intentions as they provide information about the nutritional content of menu items. They want to inform their customers about the number of milligrams of sodium or the number of calories in a specific item. They do this to help customers make decisions about menu selections that complement their personal dietary concerns.

Unfortunately, food-preparation staff in many operations cannot consistently produce menu items to the exacting standards required to specify accurate nutritional information. They must work quickly to satisfy many customers' orders, they do not have precise measuring tools, and they cannot "micromanage" production staff in ingredient selection and measuring.

For these reasons, many managers use general statements such as "low in fat" rather than specifying the specific grams of fat. Their goal is to provide general information and advice rather than to deliver specific promises.

Notice that *Exhibit 8.3* allows the manager to summarize information about the following for each ingredient in the recipe:

- Calories
- Cholesterol
- Total fat
- Sodium

The third step is to select nutrition software for the analysis. Software can be purchased inexpensively, and recipe analysis information is also available on the Internet. (Enter "recipe nutrition analysis" into a search engine. Take a simple recipe and follow the instructions to learn how these systems work.)

The fourth step is to enter recipe information into the nutrition software. If the operation uses recipe costing software, the person who updates recipes with changes in ingredient costs might enter the information. Other alternatives include the food-production manager, or someone in the purchasing office if centralized purchasing is used.

Entry of recipe ingredient information is very simple. Recipe ingredients may be entered one at a time, and the type of ingredient must be identified. For example, *corn* may list 100 or more types including sweet, white, canned, creamed style with no salt added, and creamed style that contains salt. Other alternatives include canned, whole kernel, frozen, kernels off-cob, and boiled. Once the specific ingredient is found, the amount of each ingredient (for example, one cup of corn) is also entered.

After all recipe ingredients and quantities are entered, the fifth step involves the manager learning detailed information about nutrition content. A wide range of information may be given on a per-serving basis:

- Calories
- Water
- Carbohydrates
- Protein
- Total fat (monounsaturated, polyunsaturated, and saturated)
- Cholesterol
- Dietary fiber
- Vitamins (10 different vitamins, for example)
- Minerals

While this detailed analysis may not be needed at most operations, it is available. Remember that managers can select the desired information to be included (see *Exhibit 8.3*).

Manager's Memo

Many communities are enacting legislation that requires establishments to post selected nutrition information about their menu items. Some laws apply only to restaurant chains with a specified number of establishments. However, there is concern that there will be an increased number of communities issuing an increased number of requirements for more and more operations.

Many regulations require, for example, that the number of calories and grams of saturated fat for all menu items must be listed. Locations for listing the information can include flyers, wall or counter signage, tray or place mats, menu boards, and self-order kiosks.

The last step in the process is to print the information on the menu or to use it for other purposes. Perhaps a fact sheet will be made available to inquiring customers. Providing information like this can show customers the care taken to purchase and prepare items, and suggests the approximate nutritional content.

TRUTH-IN-MENU CONCERNS

Menus are among the best customer communication and sales tools available to managers. In designing them, **truth-in-menu laws** in many locations require menu descriptions to be honest and accurate.

The concept of truth-in-menu relates to concerns that menu descriptions should give accurate information on quantity, quality, point of origin, and other factors to help customers understand what they are ordering. In other words, menu information should not deceive customers. Honest menu descriptions are good for the business because they are good for customer relations. Why would a manager want to show a menu photograph of eight large shrimp on a seafood platter if customers receive only six shrimp? Why would a manager incorrectly show that a salad contains crumbled blue cheese when many customers will notice the absence of that key ingredient?

Menus must be accurate in their selling price claims as well as in item and ingredient descriptions. For example, if a menu item is listed with a selling price of $6.95, this should be the charge. It is acceptable to indicate "Market price," but customers should be told the actual price when they ask or when they order.

If there are service charges, these should be stated on the menu. Service charges are optional, and most establishments do not have them. One relatively common use occurs when a percentage of the total meal charge is assessed on groups of a certain size or larger. Also, many lodging operations with room service have a mandatory delivery charge, and these fees should also be clearly stated.

A wide range of factors should be considered when menu descriptions are written:

- **Preparation style:** A product is not "homemade" unless it is prepared on site. Steaks should not be described as "grilled" if the markings are mechanically produced and the item is steamed.
- **Ingredients:** "Fresh shrimp" cannot be frozen shrimp and "maple syrup" cannot be maple-flavored syrup.
- **Item size:** There are legal definitions for products such as large Pacific oysters (a gallon cannot contain more than 65). Vendors may be good sources of information about descriptions for such menu items.
- **Health benefits:** There are very strict federal guidelines about nutrition claims on the menu. Menu writers cannot put creativity above accuracy, and nutrition claims are difficult for many operations to meet consistently.

OPEN FOR BUSINESS

RESTAURANT TECHNOLOGY

The Internet makes it easy to generate nutritional information about any standardized recipe with user-friendly data from the United States Department of Agriculture (USDA) or other nutrient database. Once completed, the process will not need to be repeated if the recipe does not change, in contrast to recipe costing systems that require changes whenever market prices of ingredients change.

Many of the nutrition analysis systems available on the Internet are very simple to use. In addition to restaurant and foodservice managers, they might also be of interest to persons desiring information about their own recipes for use in their personal life.

THINK ABOUT IT . . .

Many customers select nutritious items and "reward themselves" with dessert. Also, customers with a special occasion may revise nutrition concerns.

Does your concern about nutrition vary, or do you always have the same policy?

Many menus provide statements of caution about the consumption of raw or undercooked meat, fish, poultry, shellfish, or eggs. They may also warn customers about sulfate chemicals in wine, small bones in fish, and monosodium glutamate (MSG) in menu items. The menus sometimes identify these food products and disclose that eating them may increase the chance of an allergic reaction or other harm.

Truth-in-menu regulations are controlled by state and local public health departments. Local health department staff should be contacted for information applicable to a specific establishment.

Truth-in-menu laws require only that menus are truthful, but descriptions by servers also must be correct. Servers should be trained about truth-in-menu concerns so they can provide appropriate explanations. One example relates to food allergies, the topic of the next section.

ALLERGIES AND THE MENU

About 4 percent of people in the United States have food allergies. **Food allergies** occur when the body mistakes an ingredient in food, usually a protein, as harmful and creates a defense system (antibodies) to fight it. There is no known cure. The only way to prevent an allergic reaction is to avoid the food that causes it.

When a person eats a food to which he or she is allergic, reactions can begin quickly. Reactions include swelling of the lips, tongue, and throat, difficulty breathing, hives, abdominal cramps, vomiting, and diarrhea. Symptoms can range from mild to severe and even death in the worse cases, and reactions can occur a few minutes to up to two hours after eating the offending food.

Many customers do not mention their food allergies to servers. Instead, they rely on the menu for ingredient information. To help prevent food allergy incidents, managers may add a menu caution statement or post a sign prompting customers with food allergies to talk with the manager about ingredient information.*

Basics of Food Allergies

People with severe food allergies may experience **anaphylaxis**. This is a potentially life-threatening allergic reaction that can cause a drop in blood pressure, loss of consciousness, and even death.

When persons with food allergies dine away from home, they rely on service staff to provide accurate ingredient information so they can make informed decisions. Inaccurate or incomplete information puts these customers at risk

*This section is adapted from: The Food Allergy & Anaphylaxis Network. Welcoming Guests with Food Allergies. Fairfax, Va. 2001.

Manager's Memo

If truth-in-menu requirements are so important, why do some managers get away with indicating they serve a mile-high pie or have the greatest hamburger in the world? Laws do not prohibit making statements that are obviously made in jest and when most customers will realize the statement is an exaggeration.

These examples are very different from managers who inform their customers that they will be served Idaho potatoes or fresh Florida orange juice when the potatoes are not from Idaho and the orange juice is really frozen and imported.

Professional restaurant and foodservice managers strive for accuracy in their menus because they know their customers expect it. Today, there is another reason: It is the law.

THINK ABOUT IT . . .

Menu caution statements alert customers about health concerns. Some are mandated by government, such as undercooked food and foodborne illness. Others are voluntary.

As a manager, what types of voluntary caution statements would you list?

for a reaction, can end their dining experience, and may require ambulance transport to the hospital.

Education, cooperation, and teamwork are keys to safely serving a customer with food allergies. All employees must know about the issues surrounding food allergies. They must also know what to do if an allergic reaction occurs.

Although an individual can be allergic to any food, the eight food products shown in *Exhibit 8.4* account for 90 percent of all food allergy reactions: peanuts, tree nuts, fish, shellfish, milk, soy, eggs, wheat.

Exhibit 8.4

- Peanuts
- Tree nuts

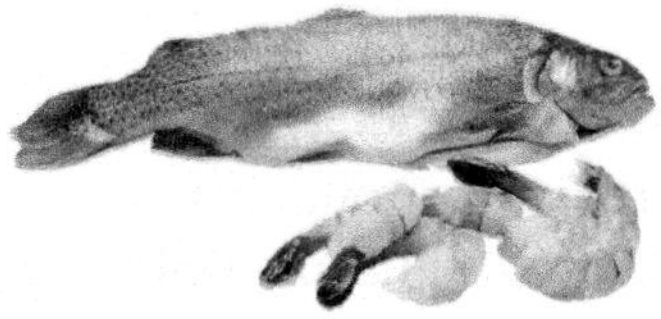

- Fish
- Shellfish

- Milk

- Soy

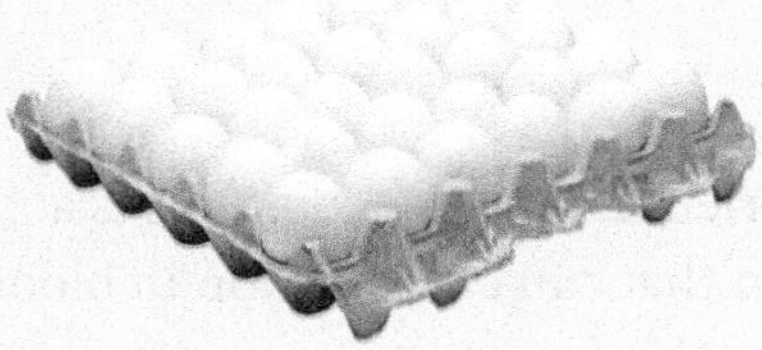

- Eggs

- Wheat

Peanuts are the leading cause of severe allergic reactions in the United States, followed by shellfish, fish, tree nuts, and eggs. Some reports suggest that fish and shellfish are likely to be the leading cause of food allergy in adults. For some people, just a trace amount of the food can cause a reaction.

Allergic reactions can occur in some persons when they consume even invisible amounts of a food. **Cross-contact** occurs when one food comes into contact with another and their proteins mix. This contact may be direct, such as consuming milk as an ingredient. It can also be indirect contact with hands or utensils, such as when a measuring tool is used for milk and then for another ingredient. Therefore, precautions must be taken to avoid cross-contact.

Manager's Memo

A study of allergic reactions to peanuts and tree nuts in restaurant and foodservice establishments showed that half of the reactions occurred because these allergens were "hidden" in menu items such as sauces, dressings, or egg rolls.

In most cases, eating the food caused the allergic reaction. Five cases did not involve consumption of the food. Two were caused by nut shells on the floor, and two involved customer contact with residual food on a table. One customer experienced a reaction after coming within two feet of the food being cooked. Cooking releases airborne proteins.

Create a Management Plan

The best way to minimize customer food allergy risks is to create a written plan. All staff members must follow the plan.

The plan should be developed around the existing menu and the recipes used to prepare the menu items. The chef should determine the menu items that are allergen free and the ingredients used to prepare the items are allergen free. The standardized recipes can be the source of the ingredients used. Manufacturers' and vendors' product fact sheets and guides from allergy associations and the USDA will be helpful, as can be the ingredient package labels. Once these allergen-free items are identified, they can be suggested to customers who indicate concerns about food allergies.

Armed with this information, the chef can consider whether ingredients known to cause allergic reactions in some people can be replaced with other ingredients. For example, the chef can confirm that the deep-frying oil does not contain peanut oil. If it does, perhaps another frying oil can be used. If the oil containing peanut products must be used, the chef will know that products fried in that oil cannot be served to those with food allergies. These items will then be placed on the allergen watch list unless they can be offered in another form such as baked instead of fried.

When creating the plan, several questions should be addressed:

- Who will answer customers' questions about menu items?
- Who will check the ingredients in menu items?
- What should kitchen employees do to prevent food from contacting other food products that frequently cause allergies?
- What should be done if a customer has an allergic reaction?

THINK ABOUT IT . . .

If a customer has an allergic reaction, it can be life threatening, cause significant disruption, and be uncomfortable for others.

Have you ever been in an establishment when a customer had an allergic reaction? What happened?

Some operations have performance standards about the timeliness of service. However, employees who feel rushed may take shortcuts. Managers should let their employees know that when they are handling food for a customer with a food allergy, more time is allowed to prevent a mistake.

Operations should be able to supply a list of ingredients when a customer with a food allergy makes this request. During operating hours, the establishment should have at least one person on duty, ideally the manager, who can handle questions and special requests. Other staff members should direct questions about food allergies to that person.

Ongoing reviews of operations may indicate how mistakes might be made when preparing a special meal. For example, can communication between the servers who take the order and the kitchen staff who prepare the food be improved (*Exhibit 8.5*)?

Exhibit 8.5

When a customer informs an employee about a food allergy, the employee should implement the plan for handling the order. That may mean providing the customer with a list of ingredients. Sometimes it means informing the customer that food items are not prepared on-site and no specific ingredient information is available.

If a customer has an allergic reaction, call emergency medical services to obtain assistance immediately. Do not delay medical treatment by denying that the reaction is occurring or waiting to see if it passes.

Post the emergency number (911) at all telephones as a reminder. Also post the street address and telephone number of the establishment by the phones.

A customer experiencing an allergic reaction should not stand. Some cases of fatal anaphylaxis have occurred after a customer has risen to an upright position. If a customer is experiencing an allergic reaction, keep him or her in the same position.

REDUCE LIABILITY

Lawsuits have been filed against operations when customers had allergic reactions. This occurred when customers were given misinformation or incomplete information about ingredients.

Ensure correct foodhandling procedures are consistently followed. Operations have been held responsible for allergic reactions resulting from cross-contact between food items after a server was notified of a food allergy. In one example, a family explained that their child had an allergy to shellfish. The child was served French fries that were prepared in the same oil used to fry shellfish. The child had an allergic reaction, and the family sued.

Compensatory damages typically cover the actual cost of medical expenses, lost wages, and compensation for pain and suffering. An operation can also be held liable for **punitive damages**. This could occur if a court found the establishment's actions showed reckless disregard for the customer's safety. Punitive damages often exceed the amount of compensatory damages.

The purpose of punitive damages is to punish offenders and serve as a warning to others not to commit the same act. Punitive damages are awarded when an individual or company was so grossly negligent that they are required to pay damages in excess of actual losses suffered. By assessing punitive damages, society makes a statement that the behavior in question absolutely will not be tolerated.

CONSIDER CUSTOMER ORDERING PROCEDURES

When a customer identifies himself or herself as having a food allergy, the employee who is informed should notify the manager on duty. Then the manager can answer questions about menu items and ensure that the proper procedures are followed.

The manager or chef should be responsible for discussing ingredient information with the customer and letting him or her know if ingredient information is not available. While employees can supply ingredients and preparation methods, the customer must decide about the specific menu selection.

Exhibit 8.6 reviews the sequence of activities for taking orders from customers with food allergies.

More about the Manager's Role

The manager should have written standard operating procedures that describe the procedures for preparing and serving food to customers with food allergies. The procedures should clearly define how to handle all communications with customers about food allergies.

Exhibit 8.6

FOOD ALLERGY ORDERS IN THE OPERATION

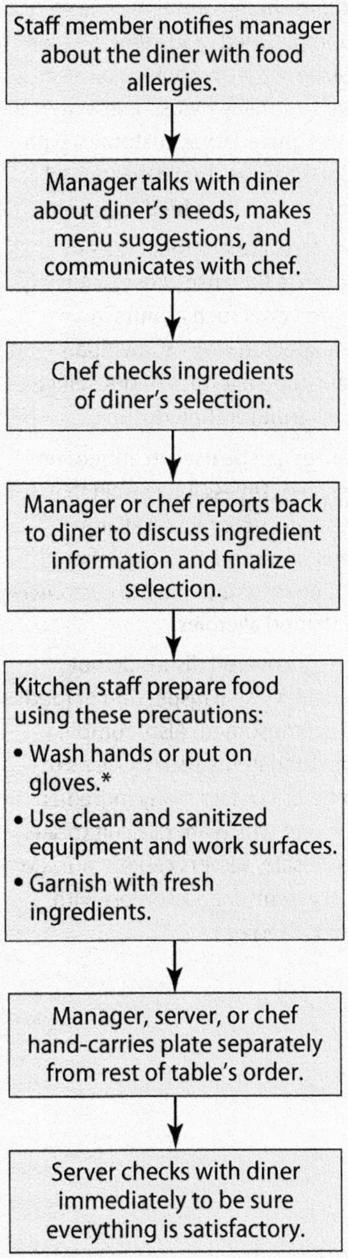

*Some individuals are allergic to latex, so this type of glove should not be used in meal preparation for such a patron.

Manager's Memo

Food allergy training should address high-risk menu choices. For example, fried food items are high risk because the cooking oil often is used for many foods. Unless there is a designated fryer, customers with food allergies should be advised against selecting fried food items.

Desserts are another concern because they use allergy-causing ingredients such as nuts in unexpected ways. Individuals who have food allergies might select fresh fruit for their dessert.

Sauces can be used in entrées or desserts. Unless it is certain that the sauces contain no allergy-causing ingredients, they should not be recommended to customers with food allergies.

Pastry-covered dishes do not permit a visual inspection of food to be consumed. Also, combination food items such as stews or pot pies contain many ingredients, some of which are difficult to see. To be safe, servers can discourage these items for customers with food allergies.

In addition, managers must develop written instructions for handling an allergic reaction. Having these plans in place before they are needed will help ensure that employees can correctly handle any allergy emergency.

Managers should also periodically conduct food allergy training to be sure that both newly hired and more experienced employees are properly prepared. This training should include a thorough review of the operation's food allergy management plan.

Buffet tables and individual service areas such as deli stations and grill areas in cafeterias are considered high risk for people with food allergies because of cross-contact possibilities. Serving utensils may be used for several dishes, or small bits of food from one dish may get into other dishes. For example, shredded cheese may wind up in a milk-free food product placed next to it. Also, buffet labels may get mixed up. If a guest with food allergies desires a specific buffet selection, consider asking kitchen employees if the item can be specially prepared.

Managers should review their menus to determine ways that customers with food allergies can learn the ingredients. Provide as much information as possible on the menu:

Instead Of	Describe As
Apple Cake	Apple-Walnut Cake
Blue Cheese Dressing	Blue Cheese and Walnut Dressing
Monterey Pasta Salad	Monterey Pasta Salad with Almonds
Chicken Stir-Fry	Chicken Cashew Stir-Fry
Asian Noodles	Asian Noodles with Peanuts
Pasta with Pesto	Pasta with Pesto (contains Pine Nuts)

If servers are not certain about all menu item ingredients, they should be trained to say so. Customers will appreciate the honesty. Servers can then speak to the chef or manager to learn details. Perhaps the chef will be asked to speak to the customer about safe menu selections.

Managers can print notes on their menus and Web sites for customers with food allergies:

- "For those with food allergies, please inform your server, who will be happy to discuss any necessary changes."
- "Please alert your server to any food allergies before ordering."

Some diners with food allergies may call ahead to find out about menu options. One person should be designated to take these calls. Managers who receive a telephone call about a food allergy–related incident should listen carefully and gather all the facts from the customer and the employees involved.

Effectively resolving customer complaints is an important part of high-quality service, so employees should be trained to not react defensively or dismiss concerns. Always view any reported allergic reaction as an opportunity to reevaluate the food allergy management plan and pinpoint areas for improvement.

More about the Servers' Role

Customers with food allergies depend on servers to notify the manager and chef about their dietary restrictions. They usually prefer not to draw unnecessary attention to themselves.

After the designated staff member has assisted the customer with the food order, that staff member should make a written notation on the guest ticket (*Exhibit 8.7*). The ticket should then be flagged at the top to alert kitchen staff about any cross-contact issues. It is not acceptable to just modify the order, for example, by writing "no cheese" for someone with a milk allergy. Instead, employees must make clear that the diner has a food allergy. One way to do this may be is to note the seating location of the allergic customer in the ordering information. However, this should supplement and not replace the need for spoken alerts according to the establishment's plan.

Exhibit 8.7

RESTAURANT TECHNOLOGY

Managers should keep all printed materials and information on the operation's Web site accurate and current to avoid confusion and reduce risks. If recipe ingredient information is on the Web site, ensure that the information is the same as the printed material available at the establishment. Posting information for customers with food allergies is worth considering because many potential customers view the Web site before visiting the establishment. Some establishments have added a link to the Food Allergy & Anaphylaxis Network Web site with a message to their customers.

KEEPING IT SAFE

Before seating a customer with a food allergy, the host should make sure the seating area has been thoroughly cleaned. This ensures that the diner will not come into contact with any leftover food from previous customers. Special care should also be taken to clean infant high chairs. In addition, a customer with a food allergy should not be seated at a table close to the kitchen's entrance. Splattering food or steam could spread allergens to the customer's table.

Special orders should not be picked up from the service area by anyone but the designated person, who might be the manager, chef, or server depending on the operation. This will help prevent problems such as delivering the wrong food or exposing the specially prepared meal to cross-contact with a food allergen. The food should be hand-carried directly to the customer. The server should ensure that nothing is accidently spilled on or brushed against the special meal.

There are several ways that allergens can enter food through cross-contact during the serving process:

- **Unclean hands or gloves:** Something as simple as picking up a muffin containing nuts and then picking up a nut-free muffin may cause cross-contact. Wash hands thoroughly and put on a fresh pair of gloves before serving an allergen-free meal. Soap and warm water are effective for removing allergens from hands, but hand sanitizers are not.
- **Splashed or spilled food:** It is possible for cross-contact to occur if a customer's food, drinks, or utensils are carried on a tray with other items. Milk or cream can spill, or butter can come in contact with a food item that is otherwise milk-free. Use a small plate or saucer when carrying cream or butter to catch any spills. Ideally, the allergen-free meal should be carried by itself directly from the kitchen to the customer.
- **Trays:** Trays used to carry allergen-free meals should first be cleaned thoroughly with hot, soapy water or other appropriate cleaning compounds and procedures. Wiping a tray with a damp towel is insufficient.
- **Garnishes:** To minimize the chance for mistakes, only the chef, manager, or other designated employee should garnish the plate. Ingredients on the production line can easily spill into containers of other ingredients. For example, it is easy for shredded cheese, croutons, or nuts to become mixed with prepped vegetables, garnishes, and herbs. To avoid cross-contact, use the backup supply of fresh ingredients.
- **Pockets:** Servers should not carry cheese graters, pepper mills, or other tools in their apron pockets.

OPEN FOR BUSINESS

KEEPING IT SAFE

What if a mistake is made on an order for a diner with a food allergy? The only acceptable correction is to discard the order and remake it. Removing the offending ingredient—for example, scraping nuts off the top of a sundae or taking the cheese off a burger—is not a solution. A trace amount of protein remaining on the food could be enough to cause an allergic reaction. If necessary, inform the customer that it will take a few extra minutes to ensure that his or her meal is prepared properly.

Exhibit 8.8

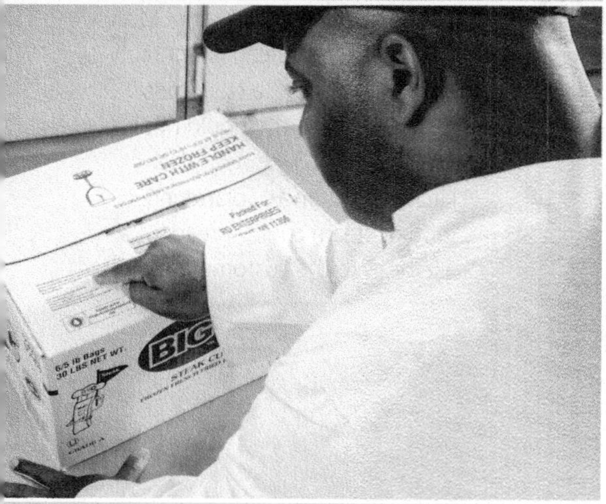

More about the Chef's Role

Customers with food allergies depend on food-production employees for answers about a meal's ingredients and cooking methods. Customers are safe only when their meal is prepared correctly.

Chefs must read ingredient and allergen information for products prepared off-site (*Exhibit 8.8*) every time they prepare a food allergy order, because manufacturers may change ingredients without notice. They should also inform diners if ingredients are used in unexpected ways. Examples include adding crushed nuts to a pie crust or using peanut butter to thicken sauces. Customers should be told about ingredients in marinades and about flavor ingredients such as butter added while cooking.

Production staff must avoid cross-contact from other dishes. There are several common ways that cross-contact can occur:

- **Unclean hands or gloves:** Wash hands thoroughly and use a fresh pair of gloves before preparing an allergen-free meal. Remember that hand sanitizers do not remove allergens.
- **Shared equipment, utensils, grills, fryers, cooking areas, and counters:** If a spatula and cookie sheet are used to prepare cookies containing peanuts, then wiped clean and reused to prepare peanut-free cookies, a customer with a peanut allergy may have a reaction. A pot of water used to boil allergen-containing food such as cheese-filled pasta may contain enough protein to contaminate other food boiled in the same water, such as milk-free pasta. Therefore, all pans should be thoroughly washed with soap and water and then sanitized. Use clean utensils for each ingredient or food item, or set aside a designated, color-coded set of utensils to handle meals for customers with food allergies.
- **Refilled serving containers:** If a container that was originally filled with cashews is then refilled with peanuts, the peanuts could have enough cashew protein on them to cause an allergic reaction. Wash and sanitize all containers carefully before refilling them with new food items.
- **Garnishes:** Ingredients on the line may spill into open containers of garnishing ingredients. Do not add garnishes to orders prepared for those with food allergies unless procedures are established and followed to prevent cross-contact. The chef or other designated staff member should apply garnishes using ingredients from the backup supply. Consider keeping garnishes that contain common allergy-causing food products such as milk, peanuts, and tree nuts in covered containers.
- **Splatter or steam from cooking:** When preparing a meal for a diner with a food allergy, do not cook the food near food containing the allergen. Do not pass other food items, plates, pans, or utensils over the pan containing the special order. Just a drop of the allergy-causing food is enough to put the customer at risk for a reaction.
- **Deep fryers:** Oil that has been used will contain protein from previously fried food. Menu items selected by a customer with allergies should be kept away from the fryer.

Manager's Memo

Chefs must read the labels of all ingredients used to prepare meals for diners with food allergies because allergens are found in some common food products:

- Worcestershire sauce contains anchovies and/or sardines; both are fish.
- Barbecue sauces vary; at least one brand contains pecans.
- Imitation butter flavor often contains milk protein, which may be listed on the ingredient statement as either artificial or natural butter flavor.
- Sweet-and-sour sauce may contain wheat and soy.
- Egg substitutes usually contain egg white.
- Canned tuna may contain casein (a milk protein) or soy protein as a natural flavoring.

If an ingredient statement is unavailable, inform the manager or customer. If possible, substitute an ingredient that is definitely safe. Otherwise, suggest another menu selection.

RESPONSIBLE SERVICE OF ALCOHOLIC BEVERAGES

Managers and employees assume legal responsibilities when they serve alcoholic beverages. They must follow all laws that apply to alcoholic beverage service. Managers must implement training and other programs to help ensure that their employees consistently follow procedures to remain in compliance. Managers are responsible for ensuring the status of any applicable licensing or certification for current employees is monitored, and that any applicable renewal requirements are met. All verification information, training records, and copies of employee beverage service licenses and certifications should be maintained in the establishment's personnel records.

Laws regulating the sales of alcoholic beverages are developed by state or the local community, and regulations may vary from place to place. However, some concerns are so important that they are addressed in legal language in every state. Two examples include serving alcohol to a minor under 21 years old, and serving customers who are, or appear to be, intoxicated.

The amount of alcohol absorbed into a person's bloodstream is called **blood alcohol content (BAC)**. In all states, it is against the law to drive a vehicle with a BAC of 0.08 or higher. This means there can be no more than 0.08 drop of alcohol present for every 1,000 drops of blood.

In addition to legal age and quantity allowed, state and local laws cover a wide range of concerns:

- Hours during which alcoholic beverages can be sold.
- Where in an establishment alcohol can be served. For example, alcoholic beverages might be served only to customers who are seated and not to those standing in a lobby area.
- Days when alcoholic beverages can be sold. For example, some jurisdictions may prohibit sales on Sunday or Election Day.
- Legal age to serve alcoholic beverages.
- Legal age to enter an establishment serving alcoholic beverages.
- Allowable "happy hours" and other drink promotions.

Alcoholic Beverages and the Body

The effects of alcohol on a customer's BAC depend on the rate at which it enters the bloodstream. The liver can remove alcohol at a rate of about one drink per hour. If consumption is greater, the BAC will increase. If, for example, a customer orders two drinks in an hour, the liver breaks down the alcohol in the first drink while the alcohol in the second drink stays in the bloodstream.

Other factors affect how high and how quickly a customer's BAC rises:

- **Drink strength:** The more alcohol in the drink, the more will end up in the bloodstream and the higher the BAC.
- **Body type:** A smaller person has a higher BAC than a larger person drinking the same amount because the smaller person has less blood to dilute the alcohol. Also, a person with a larger percentage of body fat will have a higher BAC than a lean person who drinks the same amount. Body fat does not absorb alcohol, so it must remain in the bloodstream until it is broken down by the liver. In comparison, alcohol can pass through muscle in a lean person and spread throughout the body.
- **Gender:** A woman will have a higher BAC than a man if both are of equal size and consume the same amount because women have a higher percentage of body fat. Women have a smaller amount of a stomach enzyme that helps break down alcohol. They are also typically smaller than men and have less blood.
- **Age:** An older person who drinks the same amount as a younger person will typically have a higher BAC because body fat increases with age and chemicals that help break down alcohol slow down.
- **Emotional state:** An emotional guest will have a higher BAC than a guest who is calm, all other factors being the same. When a person is stressed, angry, or afraid, the body diverts blood to the muscles and away from the stomach and small intestine. This reduced blood flow slows the absorption of alcohol into the bloodstream. The guest will not feel the effects of the alcohol and may continue to drink. As the guest begins to calm down and blood flow returns to the stomach, he or she may experience a sudden increase in BAC.
- **Medications:** Customers who consume alcohol while using many medications or illegal drugs may compound the effects of alcohol or expose themselves to dangerous interactions.
- **Food:** Customers who have not eaten have a higher BAC than customers who have eaten, all other factors being the same. Food impacts the rate that alcohol enters the bloodstream. It keeps alcohol in the stomach for a longer period of time and slows the rate at which it reaches the small intestine.
- **Carbonation:** Customers drinking a carbonated drink such as gin and tonic will have a higher BAC than those drinking a beverage without carbonation if other factors are the same. Carbonation speeds up the rate at which alcohol passes through the stomach, so the customers reach a higher BAC at a faster rate.

Additionally, some customers have a combination of these factors. For example, an elderly woman on medication consuming a carbonated beverage has four factors affecting the BAC.

Determining the Level of Intoxication

Recall that that managers and employees in establishments serving alcoholic beverages incur legal liabilities related to alcohol service. If laws are not followed, lawsuits and even criminal charges resulting in fines and even imprisonment can result. In addition, the establishment could lose its liquor license and be forced to close.

It is against the law to serve those who are obviously and visibly intoxicated. Therefore, employees must be trained to recognize when customers' levels of intoxication are increasing to the point where the service of alcoholic beverages must be slowed or stopped.

There are two ways to determine a customer's level of intoxication: count the number of drinks served and observe behavior. A combination of both approaches is best.

Proof is a measure of liquor strength and represents the percentage of alcohol in the beverage. Percentage is determined by dividing the liquor's proof by 2. For example, a 100-proof whiskey is 50 percent alcohol:

100	**÷**	**2**	**=**	**50**
Proof				**Percent alcohol**

The following beverages serve as the standard to measure when counting drinks because they all contain approximately the same amount of alcohol (½ ounce), as shown in *Exhibit 8.9*.

Exhibit 8.9

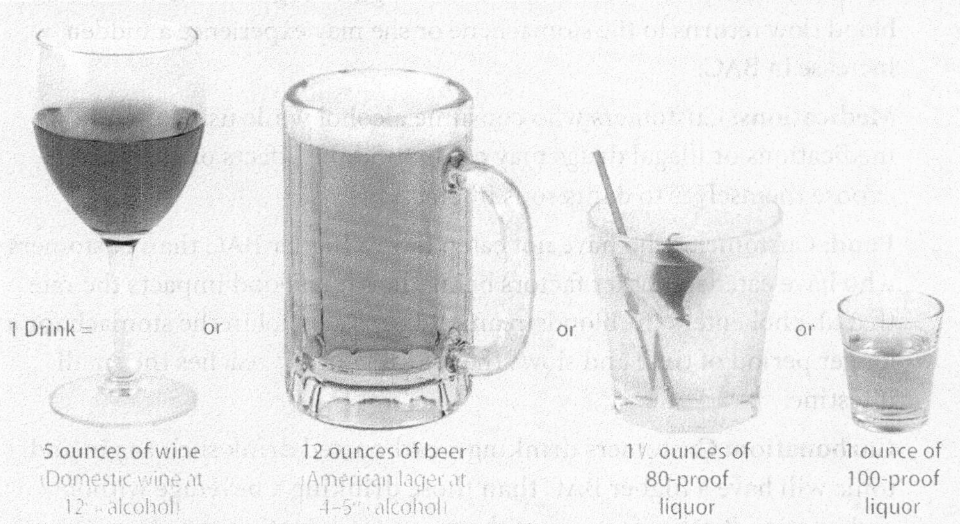

The alcohol content in a drink does not change when a non-alcoholic beverage (mixer) is added. For example, 1 ounce of 80-proof vodka contains the same amount of alcohol (40 percent) even if several ounces of mixer such as tonic are added.

If a server or bartender counts the number of drinks consumed by the customer and estimates approximate weight, it is possible to get a rough idea about that person's BAC using the information in *Exhibit 8.10* and *Exhibit 8.11*. Note that a customer's actual BAC may be different than the estimates due to other factors such as drinking before reaching the establishment or the customer's physical condition, emotional state, and consumption of food or medication. Also, some customers may exhibit signs of intoxication at lower BAC levels.

Exhibit 8.10

MEN (AFTER ONE HOUR OF DRINKING)

Number of Drinks	Body Weight 100	120	140	160	180	200	220	240
1	.022	.015	.011	.007	.005	.003	.001	.000
2	.059	.046	.038	.031	.026	.022	.018	.015
3	.097	.078	.064	.054	.046	.040	.035	.031
4	.134	.109	.091	.078	.067	.059	.052	.046
5	.172	.140	.118	.101	.088	.078	.069	.062
6	.209	.172	.145	.125	.109	.097	.086	.078
7	.247	.203	.172	.148	.130	.115	.103	.093
8	.284	.234	.198	.172	.151	.134	.120	.109

■ Indicates a BAC of .08 or higher

Adapted from: Markham, M. R., Miller, W. R. & Arciniega, L. (1993) BACCuS 2.01: Computer software for quantifying alcohol consumption. *Behavior Research Methods, Instruments, & Computers, 25*, 420–421.

Exhibit 8.11

WOMEN (AFTER ONE HOUR OF DRINKING)

Number of Drinks	Body Weight 100	120	140	160	180	200	220	240
1	.029	.022	.016	.012	.009	.006	.004	.003
2	.074	.059	.048	.040	.034	.029	.025	.022
3	.119	.097	.080	.068	.059	.052	.045	.040
4	.164	.134	.113	.096	.084	.074	.066	.059
5	.209	.172	.145	.125	.109	.097	.086	.078
6	.254	.209	.177	.153	.134	.119	.107	.097
7	.299	.247	.209	.181	.159	.142	.127	.115
8	.344	.284	.241	.209	.184	.164	.148	.134

■ Indicates a BAC of .08 or higher

Adapted from: Markham, M. R., Miller, W. R. & Arciniega, L. (1993) BACCuS 2.01: Computer software for quantifying alcohol consumption. *Behavior Research Methods, Instruments, & Computers, 25*, 420–421.

Manager's Memo

Managers should train their bartenders and service staff to start counting drinks when customers place their order and continue counting until they leave. In bar areas, a tab can be left in front of the customer so bartenders and servers can monitor it. For customers in the dining room, servers can keep a drink tally on the back of the customer check.

Counting drinks can be difficult, if not impossible, in some situations such as stand-up receptions where customers move around and are served at different locations. Then servers must be trained to rely on observation to spot possible intoxication.

THINK ABOUT IT . . .

Those who prepare and serve alcoholic beverages must learn the signs of intoxication. Why should hosts, security staff, parking attendants, and other employees also have this training?

Refer to the exhibits to see how the information can be used. A 120-pound woman has consumed two drinks in an hour. Using *Exhibit 8.11*, note that her approximate BAC would be 0.059. However, it would take four drinks for a 200-pound man to reach the same BAC (*Exhibit 8.10*).

SIGNS OF INTOXICATION

Managers should train employees to watch for physical and behavioral changes in customers. A change in behavior is more significant than the behavior itself. For example, one customer may normally be loud and outgoing, while another becomes loud after several drinks. Taking the time to talk to customers, in addition to observing, helps determine the purpose of their visit and the level of intoxication. If, for example, customers are determined to "get drunk," it is important to know this.

When large amounts of alcohol reach the brain, it cannot function normally. Certain physical and behavioral changes occur:

- **Relaxed inhibitions:** Customers with relaxed inhibitions may be overly friendly. On the other hand, they may be unfriendly or depressed, may become loud, use foul language, or make rude comments.
- **Impaired judgment:** Customers with impaired judgment may complain about the strength of a drink, begin drinking faster, or switch to larger drinks. They may make irrational statements or become careless with their money, for example, buying drinks for strangers.
- **Slowed reaction time:** Customers with slowed reaction time may talk or move slowly, lose their train of thought, or become drowsy. Their eyes may become glassy.
- **Impaired motor coordination:** Customers with impaired motor coordination may stagger, stumble or fall down, and be unable to pick up objects. They may sway when sitting or standing or may slur their speech.

Customers should be monitored from the time they arrive until they are ready to leave. If there are signs of intoxication, managers and specified coworkers must be notified. If customers move from the bar to the dining room, information about the amount of alcohol consumed should go with them.

PREVENTING INTOXICATION

Bartenders and servers must do everything possible to ensure their customers do not become intoxicated:

- Offer food to help keep alcohol in the stomach and slow the rate at which it reaches the small intestine (*Exhibit 8.12*).

Exhibit 8.12

- Offer water. Drinking alcohol causes dehydration, makes customers thirsty, and causes them to drink more. Servers can offer water with drinks and refill water glasses frequently.
- Avoid overpouring when mixing drinks. Overpouring makes it difficult to count the actual number of drinks. It also makes it difficult for customers to regulate their own drinking.
- Serve customers one drink at a time to help pace consumption.

CHECKING CUSTOMERS' IDENTIFICATION

It is against the law to serve alcoholic beverages to any customer who is less than 21 years of age. It is very important to ensure that customers are of this legal age, and this can be done only by accurately checking identifications (IDs). The types of ID that are legally acceptable vary but in most states, a driver's license, state ID card, military ID, and passport are legally acceptable.

When checking an ID, the employee must verify that it is genuine:

- Make sure the ID is valid. It must contain the owner's birth date, signature, and photo; be current; and be intact (not damaged or manipulated). Many states have training programs that explain how to determine if IDs are valid.
- Ensure that it has not been issued to a minor. All states add special features such as special colors, text, or layout features on a minor's ID to make it easy to spot.

- Use the birth date to verify customer age as shown in *Exhibit 8.13*. Many state IDs indicate the date a minor turns 21 to eliminate the need to calculate based on birth date. Some states do not provide this information, so calculations are needed.
- Verify that the ID is genuine. Servers must be familiar with valid IDs in their state and surrounding states. Fortunately, some states have developed IDs that are difficult to alter, and available ID checking guides provide full-sized samples and detailed information about minor IDs, state ID cards, and valid driver's licenses.
- Verify that the ID belongs to the customer. To do this, compare the customer to the photo. Look at physical characteristics on the ID including height and weight, eye color, and gender.

Exhibit 8.13

CHECKING AGE

A good way to determine the customer's age based on birth date involves three steps:

Step 1 Add 20 to the customer's birth year.

1990	+	20	=	2010
Customer's birth year				Total

Step 2 Add 1 to the total.

2010	+	1	=	2011
Step 1 total				Calculated year

Step 3 Compare the calculated year to the current year.

2011	to	2012
Calculated year		Current year

If the calculated year is *before* the current year, the customer is 21 years old or older.

If the calculated year occurs *after* the current year, the customer is underage.

If the calculated year *matches* the current year and the customer's birthday has passed, the customer is 21 years old. If the customer's birthday has not passed, the customer is underage.

Employees must check the ID of any customer who appears to be under 21 years old. Some establishments are very conservative and request ID for all customers, including those who are obviously much older than 21.

It is important to use proper procedures for checking IDs. The customer should be greeted properly and asked to provide an ID, and then the ID must be verified. If there are any questions about the authenticity of the ID, further verification is needed such as a second valid ID, comparison of the customer's signature to the ID signature, or asking questions such as "What is your address?" At this point, obtaining assistance from the manager or another designated employee may be helpful. If necessary, a potential customer should be refused service of alcoholic beverages. Employees have the legal right to do so in order to comply with the law.

THINK ABOUT IT . . .

If a false ID is spotted, the employee should refuse service and may refuse entry into the establishment. Always follow company policy driven by applicable laws.

If you were a manager, what would your policy be? Why?

HANDLING INTOXICATED CUSTOMERS

Even when managers train their staff to serve alcoholic beverages responsibly, problems can still arise. Establishments have different policies about who should be involved in stopping service. These policies should always be followed. In some operations, servers can do this. In others, servers must contact the manager before stopping service and in still other operations, this is a management responsibility.

Several steps are useful when stopping the service of alcoholic beverages to a customer:

- Alert a backup who is prepared to provide assistance and is close enough to observe the situation.
- Enlist the help of other customers if appropriate. Sometimes a server can ask another customer who is with the intoxicated customer for assistance.
- Wait until the customer orders the next round before stopping service. Sometimes a customer may decide that the current drink is the last, and a potential problem can be avoided.
- Inform the customer that service will be stopped. Try to do this quietly and without being judgmental. Do not say "You're drunk." Instead, say something like "I cannot serve you any more alcohol," and then express concern: "I just want to make sure you will be safe getting home." Be firm and once the decision to stop service is made, stick to it.
- Offer non-alcoholic alternatives such as coffee, soft drinks, or other beverages.

If an intoxicated customer attempts to drive away from the establishment, several strategies become important:

- Try to convince the customer that it is not safe for him or her to drive.
- Ask for the customer's keys and arrange alternate transportation. Examples include calling a person suggested by the customer, asking a companion who is not intoxicated to drive, or calling a cab (*Exhibit 8.14*). If the customer insists on driving, indicate that you will call the police—and do so. Information about the make and model of the car, the license plate number, and the direction the customer is driving will be helpful.

Exhibit 8.14

Sometimes customers arrive at the operation intoxicated. In this case, ensure they are not served any more alcohol by communicating the customer's condition to coworkers. Even though employees have not served the intoxicated customer, it is still important to keep him or her from driving away in that condition. The steps just described also apply to a customer who is intoxicated on arrival.

Manager's Memo

Designated driver programs can be helpful in the establishment's plan. For example, one person in a group of drinkers can consent to be the designated driver and not consume alcohol during the visit. In exchange, this person may be given complimentary food, non-alcoholic beverages, or coupons for future visits.

Even if a designated driver is in the group, employees should not serve another member of the group who is intoxicated. Let members of the group know that no one in the group can be "overserved." A designated driver could potentially take an intoxicated person to his or her vehicle in another location.

FOLLOW-UP ACTIVITIES

When a situation regarding alcoholic beverage consumption occurs, many establishments require that an incident report be completed. One purpose is to document what happened and what actions were taken. An incident report also helps managers determine if policies are effective or whether they need revision.

Information should be accurately reported, and the document should be completed immediately after the incident so important facts will not be forgotten. Company policy about what to include in the report and how the incident should be documented should be closely followed.

There are several times when incident reports should be completed:

- Alcohol service has been stopped to a customer.
- Alternate transportation has been arranged for a customer.
- A customer's ID has been taken.
- An illegal activity or violent situation has occurred.
- A customer has become ill.

TRAINING IS CRITICAL

Managers must train employees to effectively handle all of the potential concerns related to the responsible service of alcoholic beverages. The National Restaurant Association's ServSafe Alcohol program is one example of a system that can be presented with modification to fit the needs of the specific operation. The strategies for planning, delivering, and evaluating training programs discussed in chapter 7 will be helpful when delivering these programs. Operations that are part of a larger multiunit group may offer training programs developed by their organization for properties within the chain. Information including attendance, training date, and certificate of completion should be filed in each employee's records.

SUMMARY

1. **Explain how effective managers plan ways to incorporate a nutrition emphasis into their menus and food-preparation procedures.**

   Nutrition concerns are a long-term trend that must be addressed. The best managers consider nutrition alternatives when menus are planned and find ways to make simple recipe changes to accommodate nutrition-conscious diners.

Managers can determine the approximate nutritional content for portions produced by standardized recipes. They do so by developing a nutrition worksheet and using special recipe software to estimate caloric, cholesterol, total fat, sodium, and other desired nutritional information based on ingredients.

2. **Review examples of truth-in-menu concerns that must be addressed when menu descriptions are written.**

   Truth-in-menu laws in many locations require that menu descriptions be honest and selling prices and service charges be accurate. Examples of information that should be carefully described include preparation style, ingredients, item size, and health claims.

3. **Describe elements in an organized system for ensuring the health of customers with food allergies.**

   There is no known cure for food allergies, and the only way to prevent a potentially fatal reaction is to avoid even trace amounts of the offending food. Customers with food allergies rely on accurate menu descriptions and information from employees to learn about the ingredients in menu items they select. Most allergic reactions are caused by peanuts, tree nuts, fish and shellfish, milk, eggs, soy, and wheat.

   Managers must train employees in an organized, written plan that indicates who will answer customers' questions and who will check menu item ingredients. Kitchen employees should use care to prevent food from contacting other food items that cause allergies. Staff must recognize and know what to do if a customer has an allergic reaction. This system involves close teamwork between managers, servers, and food-preparation staff.

4. **Explain concerns about the service of alcoholic beverages.**

   Laws regulating the sale of alcoholic beverages are developed by states and local communities, but serving alcohol to someone under 21 or someone who appears to be intoxicated is illegal in all states. Persons with a BAC of 0.08 or higher cannot drive a vehicle.

   Factors that affect how high and quickly a customer's BAC rises include the amount of alcohol consumed, body type, gender, age, emotional state, medications, whether food has been consumed, and a drink's carbonation.

   Counting the number of drinks in comparison to approximate weight can provide a rough idea about a person's BAC. Signs of intoxication include relaxed inhibitions, impaired judgment, slowed reaction time, and impaired motor coordination. Checking customers' identification is important, and employees must be trained to tell if an ID is genuine and if a customer is at least 21.

   Before stopping alcoholic beverage service, a backup employee should be alerted, and it may be possible to enlist the help of other customers. When another drink is ordered, the customer should be informed that service will be stopped. If an intoxicated customer attempts to drive away, car keys should be requested. If the customer drives away, the police should be contacted. The same rules apply to a customer who arrives intoxicated.

APPLICATION EXERCISE

Learn information about legal requirements for alcoholic beverage service and consumption in your state by viewing the Web site of the Alcoholic Beverage Control (ABC) Commission. Type in "(name of state) alcoholic beverage code" or "department of liquor control." You can also try "(name of state) alcoholic beverage laws" or "alcoholic beverage control board."

When you reach the site, review the information and answer the following questions:

1. What types of liquor licenses are available in your state?
2. What are some examples of local provisions for the sale of alcoholic beverages?
3. What are the penalties for violation of liquor laws?

REVIEW YOUR LEARNING

Select the best answer for each question.

1. **In what type of operation is a dietician most likely to be employed?**
   A. Bar
   B. Hotel
   C. Hospital
   D. Restaurant or foodservice operation

2. **What is an example of a nutritional change in a menu item that has little impact on food-production staff?**
   A. Omelet made with egg whites
   B. Butter left off baked potatoes
   C. Fish served broiled instead of fried
   D. Cake with applesauce replacing fat

3. **What is an example of a menu description that could be acceptable even if it was not accurate according to truth-in-menu laws?**
   A. 4-ounce serving
   B. 1½ cup of soup
   C. 6 stuffed mushrooms
   D. Best steak in the state

4. **What is an example of a substance for which a statement of caution should be placed on a menu?**
   A. Saturated fat
   B. Sulfates
   C. Sodium
   D. Cholesterol

5. **How should a person with a food allergy prevent an allergic reaction?**
   A. Always take prescribed medicines.
   B. Get tested by a specialist regularly.
   C. Avoid food that causes problems.
   D. Eat only food prepared at home.

6. **What is the result of anaphylaxis?**
   A. Drop in blood pressure
   B. Pain in the arm or foot
   C. Loss of hearing
   D. Red spots on the arms

7. **What is the leading cause of severe allergic food reactions in the United States?**
   A. Eggs
   B. Peanuts
   C. Milk
   D. Wheat

8. **How old must persons must be to consume alcoholic beverages in all states?**
   A. 18
   B. 19
   C. 20
   D. 21

9. **At what BAC level is it illegal to drive in any state?**
   A. 8.0
   B. 0.80
   C. 0.08
   D. 0.008

10. **Why will a woman have a higher BAC than a man if other factors are equal?**
   A. Women tend to have more body fat, which affects absorption.
   B. Women have more blood to dilute the alcohol in their system.
   C. Women have more muscle than men, affecting absorption.
   D. Women react to carbonation differently than men.

9

Managing Buffets, Banquets, and Catered Events

INSIDE THIS CHAPTER

- Managing Buffets
- Managing Banquets
- Catering
- Overseeing Special Functions

CHAPTER LEARNING OBJECTIVES

After completing this chapter, you should be able to:

- **Review procedures for effectively managing buffets.**
- **Explain how to manage banquets.**
- **Describe basic procedures for managing catered events.**
- **Explain basic activities for overseeing special functions.**

KEY TERMS

banquet, p. 259

banquet agreement, p. 261

banquet event order (BEO), p. 261

buffet, p. 254

cancellation clause, p. 264

cash bar, p. 263

catering, p. 266

guarantee (banquet), p. 264

hosted bar, p. 263

sneeze guard, p. 258

CASE STUDY

"I'm happy to admit it," said Luka, the dining manager at Hawthorn Gardens Restaurant. "Customers really like the buffet. However, there is one problem we haven't worked out."

"I'm glad you're bringing this to our attention. What is it?" said Penny, the assistant manager.

Luka replied, "We agreed that the cooks would resupply the buffet during the busiest times, and we in the dining room would keep it supplied at the beginning and end of the meal period. Each team would be relieved when they are busiest."

"That's right. Isn't it working out?" asked Penny.

"Everyone is busy all the time because the buffet is so successful. No one is free to supply the line and keep it clean. I think we need to assign one person full time to that," said Luka.

1. If you were Penny, how would you decide the best way to resolve this problem?
2. How would you train an employee to keep the buffet line clean and resupplied?

MANAGING BUFFETS

A **buffet** is a style of foodservice in which customers select the menu items and portion sizes they prefer as they pass along one or more serving counters. Buffets are commonly offered in schools, hospitals, university residence halls, and business and industry cafeterias. In addition, some restaurants and many hotels offer buffets during some or all dining periods. Some buffets use food-production staff to carve roast beef or ham or to make omelets to order.

Buffets are more common, and are found in more types of operations than one might at first expect. Using the preceding definition, salad bars and dessert bars are buffets that allow customers to help themselves to one course of a meal (*Exhibit 9.1*).

Exhibit 9.1

Like all other aspects of an operation, the decision to offer a buffet requires careful thought. A significant amount of space is required in the dining area, and buffet serving equipment is expensive. Managers must consider the number of revenue-producing customers who might otherwise be seated in the space to be used for buffet service.

The location of the buffet relative to the kitchen is also important. Will it be easy for employees refilling buffet items to move from the kitchen through customer areas to the buffet's location? Is space available where customers won't disrupt others when they wait in line at the buffet?

THINK ABOUT IT . . .

Although buffet customers help themselves, employees must refill and clean the buffet. Servers are needed for setting and clearing tables.

Do you think labor costs are lower in operations with buffets and table service?

Buffet Menu Planning

Buffet menu planners have the same basic concerns as those planning menus for traditional table service. They must consider what customers want, and they will be confronted by production limitations including space, equipment, and employee skills. Additionally, tools and equipment for cooking food in batches are necessary, along with the space to store these items. Recipes that yield a relatively large number of portions must be identified, tested, and adjusted to meet the needs of the operation.

These concerns are compounded if the operation offers an à la carte menu to customers who do not choose the buffet. Other challenges driven by the menu arise if customers are served in the à la carte dining room at the same time a buffet in a banquet room is offered.

PRE-COSTING BUFFETS

How can managers price buffets when they do not know what food items and in what quantities their customers will select? One common method involves determining the average food cost for one customer from previous buffets and using this food cost as the basis to evaluate the current selling price. Then several questions can be answered: "Is the current price reasonable given our food cost?" "Should the price be increased?" "Should we carefully review the items being offered and replace some items or revise the recipes used to produce them to lower the food cost and keep the price the same?" *Exhibit 9.2* shows a worksheet that can be helpful.

Exhibit 9.2

PER-CUSTOMER FOOD COSTING SHEET FOR BUFFETS

Date: ______

(1)	(2)	(3)	(4)	(5)	(6)	(7)	(8)	(9)	(10)
Self-Service Item	**No. Portions (Beginning)**	**Refill Portions**			**Total Portions**	**Leftover Portions**	**Total Portions Served**	**Serving Cost**	**Total Item Cost**
Chicken Lasagna	24	24	24	12	84	8	76	$0.94	$71.44
								Total:	**$986.75**

$986.75	÷	**122**	=	**$8.09 (rounded)**
Total item cost (column 10)		**Number of customers**		**Per-customer food cost**

Chicken lasagna is one item offered on the buffet. It is prepared with a standardized recipe, and a pan with 24 portions is issued to the buffet at the beginning of service (column 2). During the meal period, 60 additional portions are brought to the buffet line (columns 3–5). Therefore, 84 portions were available during the meal period (column 6).

24	+	60	=	84
Beginning portions		**Refill portions**		**Total portions**

When buffet service ends, 8 portions of the item remain (column 7). Therefore, 76 portions were served (column 8):

84	−	8	=	76
Total portions		**Leftover portions**		**Portions served**

Since each portion costs $0.94 (column 9), the total food cost for all portions of chicken lasagna served on the buffet during the meal period was $71.44:

76	×	$0.94	=	$71.44
Portions served		**Per portion**		**Total cost**

The calculations just described are repeated for each item on the buffet during the meal period.

The total item cost for all portions of all items served on the buffet in *Exhibit 9.2* is calculated in column 10. This total is divided by the number of buffet customers to calculate the per-serving cost. In *Exhibit 9.2*, the total food cost is $986.75, and the total number of buffet customers is 122. Therefore, the per-serving food cost for the buffet was $8.09:

$986.75	÷	122	=	$8.09
Total food cost		**Customers**		**Per-serving food cost**

The Per-Customer Food Costing Sheet for Buffets in *Exhibit 9.2* can be completed for several days, and then the average per-customer food cost can be calculated. This cost can be used to evaluate the buffet selling price using one of the pricing methods described in chapter 1.

If a salad bar is offered along with entrée choices on an à la carte menu, the total per-customer food cost just calculated could be added to the food cost of all other choices such as vegetable and salad dressings provided with the entrée. This total food cost would be used to calculate the entrée selling cost using one of the methods in chapter 1.

The process of determining per-portion self-service food item costs does not need to be ongoing. The worksheet could be used for several meal periods and the results averaged. As product costs change over time and as the items offered are changed, the process should be repeated.

Controlling Food Costs

Managers offering self-service food items have a challenge: how to control food costs when customers decide the portion size. To address this question, managers should remember why customers like buffets:

- **Flexibility:** They can select what and how much to eat.
- **Value:** They like to receive a bargain.
- **Unlimited food:** The "all-you-can-eat" concept is a strong attraction.
- **Speed and convenience:** There is no waiting period for food production or service.

Managers of buffet operations can control food costs. Efforts should begin by recognizing that food waste is at the heart of the problem. Food waste reduces profits because it increases food costs.

Numerous strategies can be used to reduce food waste in buffet operations:

- "Learn from the garbage can." Bus staff and dish washers may know more about food waste than their managers. Which items are most frequently wasted? Offering items that customers do not like just to reduce food costs makes no sense. There is no difference whether waste is scrapped from a customer's plate or emptied from serving containers at the end of service (*Exhibit 9.3*). Higher costs result, and customers do not receive value.
- Keep track of variety and the volume of items produced and left over. Review production and leftover records to determine if changes are needed.
- Ensure that the buffet menu is flexible. If market prices increase, substitute items can be offered.
- Use standardized recipes to provide the consistency needed to please customers and manage costs.
- Consider ways to use portion control. Can items be pre-portioned into individual casserole dishes or plates of a specified size? Can chicken be cut into quarters instead of halves? Pork chops and similar items should be separated so the portions do not stick together.
- Replenish self-service stations with less food, but more frequently. Customers will perceive that food being brought from the kitchen is fresher than that on the serving line. Also, customers seeing a large volume of food may think there are "limitless" amounts and help themselves accordingly.
- Use the right serving utensils. A ladle too large for a sauce or stew makes it easy for customers to take too much and hard for them to take a smaller desired portion.

Exhibit 9.3

THINK ABOUT IT . . .

"All you can eat" might suggest a customer should see how much can be consumed, while "help yourself" is more of an invitation.

What are other simple things managers can do to reduce food costs?

- Use the correct size service items. Plates and bowls that are larger than necessary encourage customers to take more food.
- Consider the portion size of self-service items compared to their counterparts in an à la carte operation. Portion sizes are typically smaller in the buffet operation because there is a larger variety and customers can select more food if they wish. Managers can determine the proper portion size by considering the type of food, its cost, and other methods of food presentation. Employees who work in service stations should be trained in the correct portion sizes to offer.

Food Safety and Buffets

Exhibit 9.4

Customers choosing food from buffets can unknowingly put themselves and others in danger. For example, customers may eat from their plates or sample food while moving through the line. They may pick up items with their fingers or use them to taste food. They may return unwanted food items, use a soiled plate for a second helping, or put their heads under the sneeze guard to smell or to reach items in the back. A **sneeze guard** or food shield is a see-through solid barrier used to protect food in a self-service counter from customers who might cough or sneeze (*Exhibit 9.4*).

Several useful practices can be used to prevent contamination during buffet service:

- Monitor self-service areas closely using employees trained in food safety.
- Post signs with polite tips about self-service etiquette: "Please use a clean plate to help yourself to items in our salad bar."
- Maintain proper food temperatures. Always keep hot food hot at 135°F (57°C) or higher, and cold food cold at 41°F (5°C) or lower.
- Keep raw meat, seafood, and poultry separate from ready-to-eat items.
- Ensure that sneeze guards are properly installed. They should be 14 inches (36 centimeters) above the food counter, and they should extend 7 inches (18 centimeters) beyond the food.
- Identify all food items. Label containers and place salad dressing names on ladle handles.
- Do not let customers refill soiled plates or use soiled utensils. Customers can use the same glassware for refills if beverage-dispensing equipment does not come in contact with the rim or interior of the glass.
- Never use ice that is used to keep food or beverages cold as an ingredient in another food.

MANAGING BANQUETS

Banquets are events in which a sponsor pays an establishment to provide specified food and beverage services to all event attendees. They are usually held on-site, although off-site events are offered by some operations. Off-site services (catering) are discussed later in this chapter.

Unlike many hotels, restaurant and foodservice operations do not typically have separate banquet facilities. Instead, they may set aside an area of the dining room or perhaps have a separate small room normally used for groups who order from the menu.

Banquets Can Be Good Business

Owners spend a good deal of money to purchase or lease land, buildings, customer furnishings, storage, production, and service equipment and other items required to operate their business. Some managers try to maximize use of these resources to generate additional revenues without additional capital. Banquets offer the opportunity to do this. Unfortunately, banquets can also create financial problems if they are not well planned.

Banquets are often more profitable than à la carte dining. First, it is easier to schedule labor so employees are used efficiently. In a normal establishment routine, some shifts may be less busy. Managers may worry about unexpected customers and schedule more labor hours than needed. By contrast, the number of banquet customers and their arrival, dining, and leaving times are known. Employees can be cost-effectively scheduled. Banquet menus are planned well in advance so that purchasing, production, and pricing tasks are also easier.

Banquets often generate greater per-customer revenue than à la carte dining. Banquets typically celebrate special occasions, and higher-priced items can be sold. Managers also have opportunities to sell items such as alcoholic beverages to customers who might not otherwise buy them.

Profitable banquet operations begin with a careful analysis of the marketplace. Managers must learn if there is an unmet need for group dining services. Successful banquet managers have several things in common:

- They remember what their core business is. For most, it is à la carte dining. They do not take on challenges that divert resources critical for its success.
- They know when to say no. They carefully consider the pros and cons of a banquet operation to be offered in addition to à la carte dining.

THINK ABOUT IT . . .

A banquet differs from a group ordering from the menu. Banquets generally have a set menu known in advance while groups may order many different menu items.

What are other differences between banquet and group orders?

OPEN FOR BUSINESS — RESTAURANT TECHNOLOGY

Technology enables managers to hold special events such as banquets without a sponsor. For example, a manager may use data mining procedures that allow the analysis of sales records to determine other sales opportunities.

The sales records of customers who have visited the property several times during a specified number of months can be analyzed to determine which typically purchase wine with their meal. The manager can then plan a wine and food pairing banquet using these customers as the market. A special menu can be planned, special wine can be purchased, and an advertising campaign can be planned. It may be possible to find a wine distributor who will help fund part of the event.

While data mining can be done manually, the information provided by point-of-sale (POS) systems makes the task much less time-consuming and more effective.

- They understand that banquet operations involve numerous details. They commit the time required for planning, training, considering equipment purchases, and the many other activities needed to offer banquets.
- They understand that communication and teamwork are critical. Production staff must be involved in decisions about banquet menus to ensure that promises can be delivered.
- They know service or production errors, such as having too few servers or running out of food, can affect the operation's reputation. Such a mistake will be seen by many customers, not just one or two diners.
- They understand that a strong banquet business can have a positive impact on their success. Satisfied attendees may become regular customers.

Banquets Are Different

Do managers have to train their employees when they begin to offer banquets? After all, cooks know how to cook, servers know how to serve, and food preparation and serving procedures are basically the same. In fact, there are many differences, and training is needed.

First, consider food production. How will items typically prepared in small batches be prepared in larger quantities? If the banquet is not a buffet, where and how will food be plated? How will hot food be kept at the proper temperature until served, which will involve transport to the banquet dining area? Where will the transport equipment, covers for hot plates, and other items be stored?

Beverage employees also require special training. How will banquet bars be stocked? How will cash, tokens, coupons, or other beverage payments be collected? How will guest IDs be checked in a crowd of customers?

Banquet service staff must be trained. What is the necessary *mise en place* for banquet service? Will a foodservice style other than that used in à la carte dining be available? How will food be picked up from the plating area? How will it be transported to the dining area? How will soiled dishes and other tableware be removed, and where will they be taken?

Management staff will also need training. How will banquet sales be generated? What negotiation strategies should be used when interacting with banquet clients? How will information flow between managers and food-production and dining staff so no details are forgotten? What factors should be considered to determine if a specific date, banquet menu, or other aspects of a special event are possible?

Manager's Memo

All or most banquet customers will consume the same meal and will need the same serviceware. An establishment serving 300 meals nightly may need relatively few soup bowls if this menu item is not especially popular. By contrast, a banquet serving soup to 75 customers will require 75 soup bowls.

Managers are typically advised to use the same patterns of tableware in à la carte and banquet operations to reduce inventory needs. However, additional quantities may be needed. If serviceware patterns are commonly available in dealers' stocks, order times may not be an issue. By contrast, special-order patterns may require much longer lead times.

Banquet Menu Planning

Many factors considered when planning à la carte dining menus apply to planning banquet menus. There are, however, some special concerns. Menu planners must be certain that the items offered can be produced in the appropriate quantity at the desired level of quality within required time schedules. For example, it is typically easier for a broiler cook to prepare steaks as customers order them from an à la carte menu than it is to prepare many steaks to be ready for plating for a banquet all at the same time. How are the first steaks completed held at the proper temperature while others are being prepared? What about a customer that wants a steak vey well done when most persons prefer them broiled to medium doneness? Preplanned banquet menus are often offered to sponsors as suggestions. These can be helpful because they consider production limitations and desired contribution margins.

Perhaps an operation has a signature entrée, and several other items are also popular. If production equipment capacity is sufficient, these items might be logical additions to a banquet menu. By contrast, other items may be time-consuming or difficult to make in large quantities, or equipment capacities may limit their use. The manager, chef, and others on the menu planning team should identify items thought to be popular, practical, and profitable.

Managers should be willing to consider any items preferred by event sponsors. However, they must be guided by space, equipment, labor, and other limitations as they make final menu decisions.

Manager's Memo

Banquet customers almost always blame the operation for problems they encounter. This is appropriate, because most food, service, and related problems are beyond the control of the sponsor and should have been better handled by the property.

Sometimes a banquet sponsor desires menu items that cannot be produced because of equipment or other reasons. Instead of saying no, some managers unsuccessfully attempt to deliver an "impossible" menu because they want the business. The manager who made this poor decision will take the blame.

Successful managers use each banquet as an opportunity to convince attendees that the property should be the site for their next special event. They also know that successful events will encourage banquet customers to visit the à la carte dining operation.

Banquet Documents

Two important documents are needed to ensure there are no surprises as a banquet is planned. A **banquet event order (BEO)** is used by sales, production, and service staff to detail all banquet requirements. Information provided by the sponsor is summarized on the BEO and becomes the basis for the formal **banquet agreement**: a contract between the operation and the event's sponsor which specifies the responsibilities of both parties.

BANQUET EVENT ORDER (BEO)

Exhibit 9.5 summarizes the type of details included in a well-planned BEO. It identifies much of the information the operation's employees need to ensure that an event meets or exceeds the sponsor's expectations.

Exhibit 9.5

SUMMARY OF BEO INFORMATION

- General information
- ❑ Date
- Sponsor information
- ❑ Type of beverage service (hosted, cash)
- ❑ Beverages available
- ❑ Beverage charges
- ❑ Hours of beverage operation
- ❑ Menu
  - Items for each course with timing
  - Special instructions (head table and introductions)
- ❑ Room setup details and diagrams
- ❑ Table details (skirting, tablecloths, napkins, and colors)
- ❑ Music
- ❑ Audiovisual
- ❑ Coat check
- ❑ Parking

Exhibit 9.6 shows a sample BEO. Of course, each establishment will have its own form, but the information contained on the form should match the sample shown in the exhibit.

Exhibit 9.6

BOREA PRODUCTIONS BANQUET EVENT ORDER

Who and What

Event	Banquet	Guests	160	Date booked	Oct. 15, 2012
Sales	Mary Miller	Site contact	Brad Babcock	Contact	Sean Hinz
Phone	(313) 555-1217		(313) 555-0819		(313) 555-0714
Order	#2301				
Customer	Marvin Garfield	Phone	(313) 555-3258	Group	Family Services Center

When and Where

Date	July 14	Day	Sunday	Room/s	Reception in Milky Way Lobby; Dinner in Aurora Room
Times	**Start:** 5:00 p.m.	**Serve:** 6:00 p.m.	**End:** 10:00 p.m.		

Setup notes: One-hour cash bar reception with bartender

Foodservice

Menu	Quantity
Standard assorted hors d'oeuvres	4 x 120 pieces
Borealis garden salad	160
Prime NY strip steak #1180, bordelaise sauce	160
Roasted parsley potatoes	160
Steamed asparagus	160 portions
Assorted dinner rolls & butter	175
Standard chocolate mousse	160
Coffee, regular	5 gallons
Decaf	4 gallons
Wine: house red, house white	16 bottles 16 bottles

Staff	
Manager:	Mike Sellers
Chef:	Bruno Frank
Cooks:	Anna Smith, Eric Ramirez
Setup Helper:	Bryan Hansen, Dave Kapner
Setup Helper:	Cortney Breslan
Bartenders:	Bryan Hansen, Dave Kapner
Waitstaff:	Eric Baltner, Cortney Breslan, Dena Feltzies, Kevin Lopez, Ahmed Mankewicz, Jacqui Nelson, Juan Porter, Ken Pulnetzer, Alex Simmer, Joyce Tempel

Special Requests

Harpist will play at reception from 5 to 6 p.m.
Bar in Milky Way lobby from 5 to 6 p.m.
Set up tables with colored, pressed skirts
10 centerpieces

Comments

Waitstaff to arrive one hour before event to prep tables.

When they review the BEO, employees know if the event will have a **hosted bar** in which alcoholic beverages are paid for by the event's sponsor, or a **cash bar** in which alcoholic beverages are purchased by the event attendees. Other BEO information addresses room setup, linens, audiovisual, coat check, and parking requirements. Information about the menu including when the first course will be served is also noted, along with applicable billing information.

Costs for the labor needed to produce and serve banquet customers are normally included in the event's selling price. However, especially when the number of customers is small, charges for several types of labor may be assessed separately:

- Setup staff
- Bartenders
- Beverage servers
- Beverage cashiers
- Security staff
- Valet (parking) staff
- Coatroom employees
- Cleanup staff

BANQUET AGREEMENT

Exhibit 9.7 summarizes concerns in the banquet agreement. This is the formal contract between the operation and the sponsor.

Exhibit 9.7

SUMMARY OF INFORMATION IN BANQUET AGREEMENT

Contract Clause	Example
❑ Banquet room requirements	Property is not responsible for any loss of personal effects.
❑ Food and beverage requirements	Property must supply all food and beverages.
❑ Guarantee	Required 7 days in advance with final allowance (+ or − 10 %) within 2 days. One hundred sixty customers are guaranteed one week before the event. The final count, which is the basis for customer payment, is needed two days before the event. That number must be + or − 5 customers (155–165 customers).
❑ Billing or cancellation	Timing for payment (usually in advance without an approved credit application).
❑ Audiovisual equipment	Equipment provided by client is the client's responsibility.

The banquet agreement is a legal contract, so it must carefully identify responsibilities of both property and sponsor. This can require a document several pages long.

Banquet agreements can be drawn up many months or even years in advance of the event. Therefore, it is not always possible to specify menu and other prices because of possible cost changes.

Exhibit 9.7 indicates that guarantee information is important. A **guarantee** is an agreement about the number of meals to be provided at a banquet. Typically, it must be made a specified number of days in advance. At that time the sponsor agrees to pay for either the actual number of customers served, or the number of customers guaranteed, whichever is greater.

Managers make a commitment to hold space for events under contract. They must protect themselves if, for example, an event is cancelled. The space could have been used for another event. This concern is usually addressed with a **cancellation clause** that indicates any financial penalties for cancellation at certain times before the banquet date.

As with other legal documents, managers should seek the services of a qualified attorney to draw up and review banquet agreements before use.

Exhibit 9.8

Banquet Food Production

Is there any difference in how a cook prepares a portioned entrée such as a steak filet for an à la carte dining guest and for a banquet guest? In fact, isn't it easier to produce many servings of an item for a banquet instead of several portions of many items in an à la carte dining operation (*Exhibit 9.8*)?

When possible, the same recipes would probably be used in à la carte and banquet operations so items of the same quality prepared with the same ingredients would be served. However, the quantities of ingredients and portion yields might differ. Surprisingly, food production for a banquet operation in a typical establishment will not likely be easier than for its à la carte operation. Why? Because the operation's kitchen has probably been designed to produce small volumes of numerous items rather than large volumes of specific items at one time.

Consider a typical hot food production line with a range oven, grill, deep fryer, and broiler. This equipment may be appropriate for the variety and volume of items produced for the à la carte dining room. However, it may not be adequate to produce 150 portions of steak or seafood filet at one time for a banquet.

The production problem is greatly increased if a large number of à la carte meals are needed at the time of banquet production. This situation could occur when careful planning and analysis did not occur before the decision to accept the banquet. The possible benefits of several thousand dollars in additional revenue will likely be more than offset by significant negative word-of-mouth advertising that could lead to a reduction in the à la carte dining business.

Another production consideration relates to portioning. It is one thing to place a grilled steak, baked potato, and vegetable on a single plate for immediate pickup by a server. It is another thing to pre-portion many plates at one time for banquet service.

Where can one find the space required for this "assembly line"? How many employees will be needed? What type of equipment is required? Space, staff, equipment, and supplies for this specialized task are unavailable in many operations.

Banquet Beverage Production

Alcoholic beverages can be served at bars in the banquet dining area before, during, and after the meal. Wine may be served with the meal.

There are three basic pricing plans when the sponsor pays for the drinks:

- Charge by the drink. The sponsor pays for beverages consumed. The drink charge may be the normal price or another charge that is negotiated.
- Charge by the bottle. The sponsor pays for drinks based on the number of bottles issued to the banquet bar at a negotiated price per bottle. In the event of partial bottle use, there should be agreement in advance about whether the charge will be on a whole or partial bottle basis. It is important to note that in nearly every liquor jurisdiction, it is illegal for an establishment with a commercial (versus retail) liquor license to allow customers to take home opened bottles, even if they have been paid for.
- Charge by the hour. The sponsor pays a specified price per hour (or partial hour) that the bar is open for service. The manager bases the hourly rate on the number of customers of legal drinking age multiplied by the estimated number of drinks each will consume per hour multiplied by the negotiated per-drink price.

Sometimes a sponsor hosts the bar for only part of the event. Then the host bar becomes a cash bar, and customers pay for drinks. When a cash bar is used, a manager must develop procedures to ensure that all drinks produced are paid for:

- Bartenders may use an electronic register in the same way it is used in the à la carte operation. Customers pay as they are served.
- Bar cashiers may sell tickets that customers "trade" for a beverage. Different colored tickets may represent different selling prices.

REAL MANAGER

QUALITY COUNTS

After 20 years in various food and beverage management positions, I opened my own banquet and catering facility. Despite the fact that our opening occurred 2 weeks after the tragedy of 9/11, we became profitable the second year in business. We continued to grow even in the tough economy that followed, largely because we held ourselves to high standards in service and products. One of the many testaments to our commitment was a client who had four daughters who left their wedding decorations with us to use at the next daughter's wedding, which came 2 years later. He said, "Why should I take these decorations when all my daughters will be having their receptions here?" This is one of the many examples of the importance of satisfying guests through excellent execution.

Banquet Service

The best banquet service begins with a line-up meeting. This meeting tells servers about the purpose of the event, the menu, program details, and necessary instructions.

Many à la carte dining procedures apply to banquet operations. These include how place settings are arranged and basic food and beverage service methods. American service, in which food items are pre-portioned onto plates in the kitchen for service to customers in the dining area, is common in both à la carte and banquet operations.

Sometimes service styles may differ between courses. If service employees are trained appropriately, the customers may enjoy this. Consider the use of American plated service throughout the meal, and then having servers present a tray with selections of whipped cream, chocolate shavings, and cinnamon sticks for coffee service. How about a Caesar salad prepared at the head table for special customers with pre-plated Caesar salads then brought to the other customers? Buffets or self-service are also popular for banquet functions.

À la carte dining employees can be excellent banquet servers with additional training. Many operations use an "on-call" system of persons available for banquet events. Since they will represent the operation, these part-time staff members must have the appropriate knowledge and skills before serving banquet customers.

CATERING

Some operations offer catering in addition to à la carte dining and on-site banquet operations.

Catering involves the production and service of food and beverages at a location other than the establishment responsible for the event. Since it is difficult or impossible to return to the kitchen for products, supplies, or equipment, careful planning is very important to ensure everything will be available for this off-site event.

Overview

Like banquet operations, off-site catering can generate significant revenues for operations that do it well. Also like banquet operations, the sponsor selects the menu, invites the customers, and establishes the time and type of service. Unlike a typical banquet, however, the host can also select the event's location. This is an advantage to the host and a potential disadvantage to the manager. It creates a significant challenge and requires detailed planning to ensure the event's success.

There are two primary catering markets:

- **Business groups:** The opening of a new business, a retirement, an appreciation or other recognition event, and special occasions such as summer picnics and holiday parties can all be held at the sponsoring organization's location. Breakfast and lunch meetings at the client's location are also catering opportunities.
- **Individual social events:** Birthdays, graduations, weddings, anniversaries, and reunions are examples of special times that can be celebrated in locations other than a banquet room. In many areas, there is significant competition for these events, and a caterer's reputation does much to "sell" a potential client.

Political, religious, and charitable groups sponsoring fund-raising events also commonly use off-site caterers.

OFF-SITE LOCATIONS

The location of an off-site event can present a significant challenge. Many sites provide opportunities for creative caterers to use their professional talents. At the same time, meal service can be made more difficult because of restrictions imposed by the customer.

Some off-site locations such as schools and churches have adequate water, gas, and electric utilities along with storage, preparation, and service facilities. Other locations such as parks or beaches may have no facilities at all. These catering locations pose special foodhandling concerns that must be addressed, including procedures and equipment for handwashing and refrigeration.

It is critical that the caterer visit the proposed site if he or she has not had prior experience with it. *Exhibit 9.9* shows the factors that should be assessed during a visit to a potential catering location.

The caterer should determine the responsibility for usage fees for the site. Off-site locations may be arranged by the sponsor, but this is an important point to confirm. If equipment and other items not owned by the operation are to be delivered to the site, arrangements for the deliveries must be made. The caterer must know when the equipment will be delivered, where it will be stored, and how it will be secured.

If the location has usable preparation equipment and space, raw food deliveries may be arranged directly to the site to reduce the number of items transported from the operation. However, arrangements must be made for their delivery, receipt, and secure storage.

Exhibit 9.9

SUMMARY CHECKLIST OF FACTORS FOR OFF-SITE CATERING INSPECTION

- ❑ Access
- ❑ Available storage and production equipment
- ❑ Available utilities
- ❑ Dining needs:
  - ❑ Available dishes, flatware, and glassware
  - ❑ Tables and chairs
- ❑ Alcoholic beverage service (license?)
- ❑ Restrictions
- ❑ Bar needs
- ❑ AV needs
- ❑ Dance floor
- ❑ Parking
- ❑ Restrooms
  - ❑ Customers
  - ❑ Employees
- ❑ Music or band
- ❑ Decorations
- ❑ Spaces for buffet and dining tables

Managers must often be very creative in locations with inadequate food-preparation facilities. A most important concern is to plan a menu that recognizes almost everything must be produced back at the caterer's location. Experienced managers with access to water, tabletops, and an electrical outlet for microwave ovens at an off-site location can do much to supplement food preparation done at their kitchen.

Special concern must be given to the proposed dining space. If there is a scenic view or other attraction that will affect table placement, this should be noted. Tables and chairs, if available for use, should be checked to ensure that they are safe and will meet the customers' needs. They may also need to be cleaned.

Arrangements for and restrictions about the consumption of alcoholic beverages must be determined. Laws in some areas may prohibit locations with on-premise liquor licenses to allow liquor to be brought in. In other instances, the customer may purchase the liquor, and the operation provides bartenders, beverage servers, and serving supplies including glassware.

A careful inspection of the proposed location is absolutely critical for a successful event, and no "shortcuts" should be taken. Even if a location has been used recently, an inspection may be needed to ensure that previous experience applies to the event being planned.

Planning and Managing Catered Events

A series of steps should be used to plan and carry out an effective off-site catered event. This section takes a close look at each of them.

Exhibit 9.10

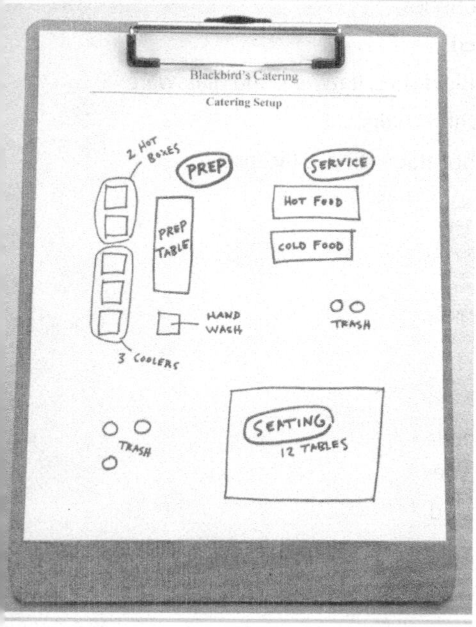

BEFORE THE EVENT

Step 1: Conduct the first meeting with the potential sponsor.

At this time, the manager can learn about the event including its purpose and location, estimated number of attendees, menu ideas, and alcoholic beverage concerns. If there is general agreement about factors important to the sponsor and the caterer, a reservation for the date is made. A deposit may be made to hold the date.

Step 2: Make the first site visit.

If the manager is not familiar with the proposed location, a visit is needed. The off-site inspection addressing the factors reviewed in *Exhibit 9.9* will suggest important concerns. A drawn-to-scale sketch of the dining area may be helpful to plan placement of serving lines, portable bars, and dining tables (*Exhibit 9.10*). During this trip, the manager may be able to estimate the transport time between the preparation kitchen and the event's location.

Step 3: Make menu planning and costing decisions.

The next step is to draft a menu that can be produced with the equipment available at the operation and the event's location. Menu planners for an off-site catered event have similar limitations to those planning a hotel room service menu: The quality of food is lowered as time goes by. This places a significant constraint on items that can be served if they must be produced on-site and delivered. This challenge may be lessened if the off-site facility has some food-preparation and holding equipment. After a tentative menu is planned, product cost decisions can be made that, in turn, impact the price quotation. The time needed for food preparation and transport can also be estimated, as can service costs.

Step 4: Meet with the potential sponsor.

At this time, costs can be negotiated and changes in the plans, if any, can be made. All contract issues should be discussed, and the event agreement should be signed by both parties. The sponsor will typically now be required to make a deposit for food and beverage purchases, especially if there are special products that must be purchased in advance.

Step 5: Plan all of the details.

Catered events are typically "simple but complicated." They are simple because the same basic planning must be undertaken regardless of the event's size. They are complicated because of the many details that must be considered at every step in the planning and delivery process. Catered events involving thousands of meals (or more) can require at least one year to plan. Smaller events with which the manager has extensive experience can be planned in several weeks. Whatever the size, the manager must be creative, concerned about details, and alert to ideas for pleasing the customers while effectively managing costs.

Step 6: Obtain the customer guarantee.

At the time specified by the event agreement, the sponsor should finalize the number of customers. A guarantee deposit is normally paid at this time. The guarantee helps in exactly the same way that it does with banquet planning: It provides the information that drives food-production and staffing plans. With the guarantee in place, the manager is assured to receive payment based on the higher of the guaranteed number or actual number of customers served. In many cases, caterers plan to serve a few more customers than the guarantee. For example, a caterer planning to serve a group with a guarantee of 200 may prepare for 210 customers in case a few more people arrive.

OPEN FOR BUSINESS

KEEPING IT SAFE

Those providing off-site foodservices must follow the same food safety rules as when they produce food for on-site consumption. Food must be protected from contamination and time–temperature abuse. Facilities must be cleaned and sanitary. Food must be prepared and served safely, and employees must follow good personal hygiene practices.

If power or running water is not available in a location, foodhandling procedures must be changed:

- Portable handwashing facilities must be set up.
- Use insulated containers to hold TCS food products. These items require time–temperature control to prevent the growth of microorganisms. For example, raw meat should be wrapped and stored on ice. Milk and dairy products should be delivered in a refrigerated vehicle or on ice.
- Store raw and ready-to-eat products separately.
- If leftovers are given to customers, provide instructions about how to handle the food.
- Place garbage containers away from food-preparation and serving areas.

DAY OF EVENT

Very large catered events may require on-site work several days or longer before the event. Perhaps landscaping must be provided or tents must be put up. Often the sponsor employs vendors to do this, and the establishment manager's responsibilities are limited to food and beverage related activities. Each event and situation will require a unique solution.

On the event day, some work must be done at the establishment. This includes food pre-preparation and preparation along with kitchen cleanup. Also, the food, beverages, equipment, utensils, supplies, and numerous other items needed must be counted and loaded on the transport vehicle(s). This is where attention to details is a must.

The off-site catering inspection form described in *Exhibit 9.9* can help remind the manager what must be transported. Packing sheets that identify everything to be shipped must be checked and rechecked, especially when different persons pack equipment, food, and supplies. If washable dishes and flatware will be used, these also must be transported unless they are available at the off-site location.

Providing food to an off-site location can present challenges when it comes to preventing time-temperature abuse of the food being delivered. During transport, sanitation concerns such as the use of proper insulated chests (*Exhibit 9.11*) are important to minimize the time that potentially hazardous food items are in the temperature danger zone.

Safety concerns are also important. Heavy transport equipment can move during transport and cause damage or injury to persons, the delivery vehicle, and its contents. Items must be carefully packed to reduce damage—gelatin molds or tiered cakes, for example, and breakable items such as dishes. Location on the vehicle is also important. Light food items, for example, should be kept together and away from heavy nonfood items.

Exhibit 9.11

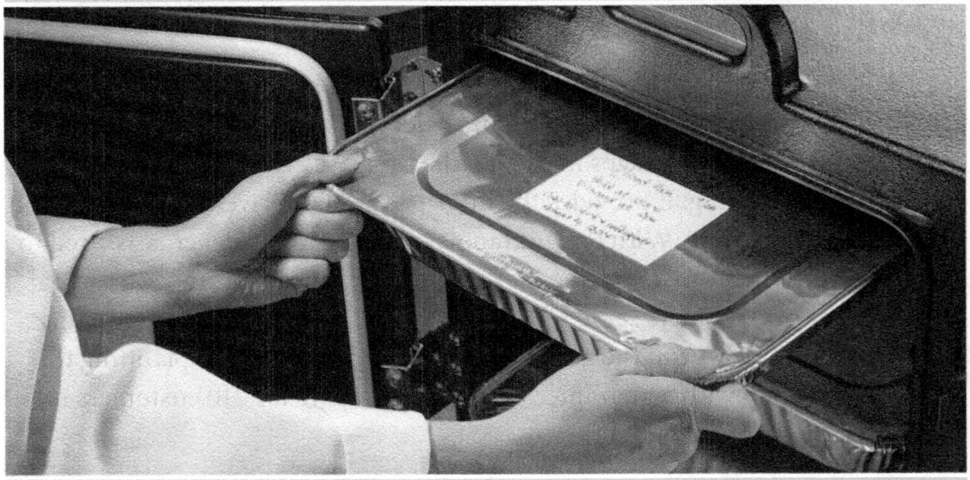

At the off-site location, food items may need to be pre-prepared, if applicable, and held at the proper temperature until serving. The dining area must be set up, food and beverage products must be served, and the location must be cleaned at the end of the event.

Typically, any remaining sponsor payments are expected at the end of the event, and all items that belong to the operation must be transported back to the property. Sometimes even the garbage from the event must be removed from the off-site location.

Back at the establishment, items used off-site must be counted, cleaned, and replaced into inventory, if applicable. Accounting activities such as preparing the sponsor's payment for bank deposit and maintaining payroll records are also important.

STAFFING THE CATERED EVENT

Exhibit 9.12 shows how the staffing for a catered event might be organized. Most roles are the same as those done in other types of food and beverage service. A chef working with assistants prepares the food, and a utility person washes the pots and pans. He or she will also need to clean washable serviceware, if used, when it is returned from the event.

Manager's Memo

Food, beverages, and all necessary equipment, utensils, serviceware, supplies, and other items must be transported to the off-site location. Most typically this is done by a manager or employee who will be working at the off-site event. However, the employees who will be working at the location must also get there. They may meet at the operation and be driven to the event site. At other times they are given driving directions and a travel expense allowance and are personally responsible for arriving at the location at the specified time.

Exhibit 9.12

STAFFING FOR OFF-SITE CATERED EVENT

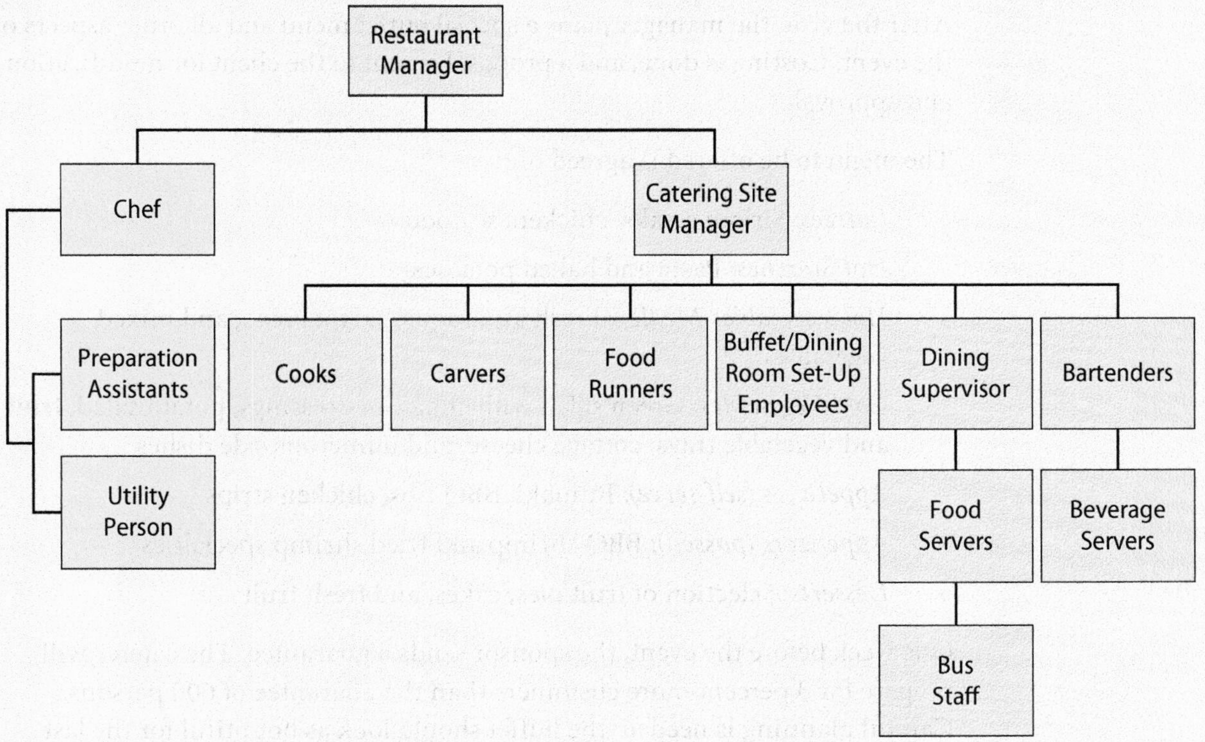

Cooks portion food items into transport containers. With help from the utility person, they also load transport equipment into the delivery vehicle.

At the off-site location, the catering site manager must ensure that all contractual terms are complied with and that the event is successful. He or she also supervises the production and service employees. Food servers and beverage staff typically do the setup and cleanup work for small events.

A line-up meeting should be conducted after all staff are assembled at the off-site location. They should be briefed about the event, the menu, serving techniques, and work assignments. The manager should use proper supervisory principles to help ensure that employees work according to standard procedures. Managers should also be empowered to make on-the-spot decisions to make the event as enjoyable as possible for the customers.

Catered Event Case Study

A real-world case study can show many of the details required to make an event successful. Consider the following example:

Six months before the planned event, the manager meets with the client. This client is a committee representing a group sponsoring a once-a-year auction fund raiser. During the meeting, tentative plans about a theme and potential menus are discussed. A tour of the kitchen in the facility owned by the event's sponsor is conducted to assess available equipment and layout and to test-operate needed equipment.

After the visit, the manager plans a special buffet menu and all other aspects of the event. Costing is done, and a proposal is sent to the client for modification and approval.

The menu to be offered is agreed on:

Entrées: Sirloin steaks, chicken, seafood

Hot Starches: Pasta and baked potatoes

Hot Vegetables Medley: Fresh asparagus, carrot coins, and mixed vegetables

Cold Vegetables: Green salad with choice of dressings, potato salad, fruit and vegetable trays, cottage cheese, and numerous side dishes

Appetizers (self-serve): Rumaki, BBQ ribs, chicken strips

Appetizers (passed): BBQ shrimp and fried shrimp specialties

Desserts: Selection of fruit pies, cakes, and fresh fruit

One week before the event, the sponsor sends a guarantee. The caterer will prepare for 3 percent more customers than the guarantee of 600 persons. Careful planning is needed; the buffet should look as bountiful for the last person served as it does for the first. The caterer will also provide food for the approximately 125 volunteer members of the organization who will help with the event. Also, approximately 35 staff members will consume at least one employee meal each. The sponsor pays a deposit of approximately 25 percent of the estimated bill, and is invoiced for the remaining amount. This remainder is to be paid at the end of the month.

All the food is ordered and delivered to the caterer's kitchen. Food preparation begins several days in advance. For example, some entrée sauces can be prepared. Appetizers such as rumaki (water chestnuts wrapped with sliced bacon) can be assembled and placed in trays that are tightly wrapped before refrigeration. To suggest the volume of items to be transported to the serving site, a typical customer consumes four rumaki appetizers. This means that about 2,400 must be prepared.

600	×	**4**	=	**2,400**
Customers		**Appetizers each**		**Appetizers**

If approximately 120 pieces can be placed in one 12" × 20" × 2" steam table pan, 20 pans will need to be transported for just this appetizer:

2,400	÷	**120**	=	**20**
Appetizers		**Appetizers per pan**		**Pans needed**

Consider also that a standard convection oven holds 10 of these pans and that it takes 20 minutes to cook the items. This means that one oven will be in use for 40 minutes:

20	÷	**10**	=	**2**	×	**20**	=	**40**
Pans		**Pans per oven load**		**Oven loads**		**Minutes**		**Minutes**

Most preparation occurs during the last two days and must be coordinated with the ongoing à la carte business. The dining room averages 225 total meals per day over midday and evening meal periods.

The caterer brings almost everything that is needed for the event to the site, including a refrigerated truck that will serve as the refrigerator. Only the off-site facility's preparation and serving line equipment will be used.

All serving line pans and utensils, dinnerware, and other customer-related service items including alcoholic beverages are shipped to the facility one day before the event. The food is loaded onto the refrigerated truck late on the evening before and is transported to the site early on the day of the event. The banquet chef personally packs or supervises the loading of all food and needed supplies and equipment following a very detailed, four-page checklist of necessary items. The chef must also ensure that items are loaded correctly and fastened securely to avoid spillage or damage during the trip.

As soon as the truck arrives, food and equipment are checked in according to the same checklist, and food preparation begins. Dining-room and serving line setup is done by employees assigned to these duties.

The self-service stations must be set up, and all serviceware for the passed appetizer service must be organized. Hot food preparation and finishing activities last all day. Staff assigned for cold food items and salads arrive in the early afternoon and begin work.

During the event, customers can help themselves to appetizers at a special station. Other appetizers are offered by food servers who circulate throughout the serving, dining, and auction areas (*Exhibit 9.13*).

The hot and cold food serving lines are set up so customers can pass down both sides. Washable china serviceware is used. It takes approximately one hour to serve the customers.

Exhibit 9.13

Volunteers from the sponsoring organization pre-wash the caterer's dishes and other serviceware before they are boxed for shipment back to the establishment. The caterer brings its own smallwares such as pots and pans and other cooking utensils, as well as a staff member to wash and account for them before they are shipped back.

Was the fund raiser a success? When the event concluded, members of the host committee said "Thank you very much, and we will certainly see you next year!" The establishment manager knew there was an excellent chance for repeat business.

OVERSEEING SPECIAL FUNCTIONS

Managers who oversee buffets, banquets, and catered events as a routine part of their responsibilities have likely resolved many potential challenges. Just as in à la carte operations, routines become established as management and decision-making concerns are addressed.

By contrast, managers who offer these services on an infrequent basis may be confronted with challenges because their experience is more limited. They must attempt to effectively plan for problems that are likely to arise and determine whether buffets, banquets, or catered events should even be offered.

The success of many special functions can be traced back to the operation's contract with the sponsor. All details including payment arrangements should be made in advance so neither the manager nor the sponsor will be surprised about any aspect of the "who, what, when, and where" of the event.

It was noted earlier in this chapter that staff assigned to special functions will likely need training even though they have experience in à la carte operations. Managers must consider every detail of every task and ensure that employees have the knowledge and skills to perform them during the event. Experienced managers know that the time to practice these activities is before, not during, the events.

Line-up sessions should be used for any à la carte or special function to update staff about details including the menu, serving times, special customers, and other details. Perhaps the best question to ask is "What must employees know to make this special function a success?" The answers to this question become the information that must be presented in a line-up session. Many of these special details are likely noted on the banquet event order.

Managers must supervise staff during the special function. As with any other type of supervisory responsibility, this task is much easier when performance standards have been established and employees have been trained to follow basic procedures.

When banquets or catered events are not routine activities, it is important that all employees know what they are to do and when. In some operations, special events are staffed by volunteer paid employees selected by managers. These employees will likely be those who are knowledgeable and skilled and who consistently demonstrate an attitude of caring about the customers and the operation.

Managers must be able to resolve unanticipated problems that arise during the special event. The knowledge and problem-solving skills necessary are the same ones useful in à la carte operations. Failure to follow plans, misunderstandings about responsibilities, and even planning errors can arise when special functions are not routine.

THINK ABOUT IT . . .

Managers with special event experience know that large groups are different from small groups.

What activities would cooks and servers need to do differently when preparing for, delivering, and cleaning up after a group function?

A monitoring system should be in place so problems can be detected early. Managers are one step behind in addressing challenges when they must be informed by the event's sponsor. A manager's ongoing responsibility is to deliver what is in the contract. One incentive to do this is the knowledge that those attending the special function are likely to tell their friends about the experience and consider the operation for their own events.

Payment requirements for the special function will have been detailed in the agreement. Concerns about amounts, payment schedule, and any other details should be addressed in that agreement. This contract then becomes the benchmark against which payments should be made.

SUMMARY

1. **Review procedures for effectively managing buffets.**

   A buffet allows customers to select desired menu items and portion sizes as they pass along one or more serving counters. Buffet menu planners must consider what customers want along with space, equipment, and employee skill limitations.

   Buffets can be pre-costed by estimating each item's portions consumed and multiplying that number by the portion cost. Totaling these results and dividing by the number of customers gives the buffet's per-serving food cost.

   Buffet food cost can be controlled if managers can reduce food waste. Strategies include making it possible for customers to select smaller portions and tracking item popularity. Food safety concerns are important, and the principles used for other styles of foodservice apply. Special concerns involve maintaining food at the right temperature and using sneeze guards to ensure food does not become contaminated.

2. **Explain how to manage banquets.**

   Banquets are events in which a sponsor pays an establishment to provide food and beverage services, usually on site. They can be good business if they meet quality requirements and do not overtax resources. Banquets generate additional revenues without significant additional capital. They do require additional training.

   Banquet menu planning concerns are similar to those for à la carte operations. Differences can include equipment limitations and the use of food products that retain quality during plating and transport to the service area.

   Banquet event orders (BEOs) detail all banquet requirements for employees. Banquet agreements are contracts between the operation and the sponsor. The sponsor must guarantee the number of customers, and a cancellation clause indicates financial penalties for cancellation at specified times.

   There are three basic pricing plans for a hosted bar: charge by the drink, charge by the bottle, and charge by the hour. A cash bar requires a method of tracking payments.

3. **Describe basic procedures for managing catered events.**

   Catering involves the production and service of food and beverages at an off-site location. It can generate significant levels of revenues, but attention to numerous details is required for successful events.

   The event's location is a special concern. A detailed checklist can confirm the resources available and what must be supplied by the operation.

   Before-event activities include meetings and site visits as well as menu planning and costing decisions. Numerous production and delivery concerns are important as is the contract specifying the sponsor's and the operation's responsibilities.

   A catering site manager is typically responsible for all food and beverage production, service, setup, and cleanup at the off-site location. Another person will be responsible for food production at the establishment and loading the delivery vehicle.

4. **Explain basic activities for overseeing special functions.**

   Managers who oversee special functions routinely typically have fewer challenges than those who provide these services less frequently. Detailed information should be supplied to all employees in line-up sessions. Managers must supervise employees during the function. This responsibility is easier when performance standards have been established and employees trained.

   Managers sometimes need to resolve unanticipated problems. Many of these can be traced to planning errors, failure to follow plans, and misunderstandings about task responsibilities. Monitoring systems should be used to identify and resolve problems promptly.

APPLICATION EXERCISE

"Something is wrong here," said Anitra, the owner of five of the restaurants in the Meadowlands Restaurant chain. She was talking to Blaine, the unit manager.

"What do you think is wrong?" asked Blaine in response.

"Well," replied Anitra, "all of our restaurants are designed very much the same with an identical à la carte menu, a versatile kitchen, and our 'Summit' meeting room that allows us to offer up to 75 person banquets. We think this is a competitive edge over our other restaurant competitors, and the banquet business should be—and is—very profitable at our other four restaurants.

I have noticed that your unit sells fewer banquet events, that they have a higher food and labor cost, and that they are much less profitable than the other units. Also, there seem to be more banquet sponsor complaints.

I am scheduling Alice, our unit manager in Pleasanton (a community about 50 miles away), to spend a week here. I want you two to figure out how to correct the problem. This property needs to have a much better-performing banquet operation."

Assume you are Alice and you are preparing a list of what you think should be happening at Blaine's property if it was enjoying a successful banquet business. Your plan is to then compare the list of best practices with the actual procedures that are used. Address the following questions related to a banquet procedures list.

1. List five procedures for meeting with potential banquet sponsors.
2. List five concerns that should be considered as the preplanned banquet menus are developed.
3. List three suggestive selling options related to alcoholic beverages for banquet events.
4. List six things that the on-duty manager should do during the event to ensure that customer service is as it should be.
5. What feedback is important from the sponsor after the event? How should the feedback be gathered?

REVIEW YOUR LEARNING

Select the best answer for each question.

1. To cost items on a buffet, the total number of portions consumed must be multiplied by the

A. standardized recipe's portion cost.
B. number of customers served.
C. number of leftover portions.
D. number of portions issued.

2. One strategy for reducing food costs in buffets is to replenish self-service stations with

A. more food more frequently.
B. less food more frequently.
C. more food less frequently.
D. less food less frequently.

3. Sneeze guards should be placed how many inches above the food counter?

A. 14 inches (36 cm)
B. 16 inches (41 cm)
C. 18 inches (46 cm)
D. 20 inches (51 cm)

4. What is a reason that banquets are often more profitable than à la carte dining?

A. More types of convenience food are generally used.
B. Food costs per portion are generally lower.
C. Fewer managers and supervisors are necessary.
D. Labor can be more accurately scheduled.

5. What is a benefit to the use of preplanned banquet menus?

A. Reduce the need to purchase food products.
B. Sell menus with desired contribution margins.
C. Reduce the time beverage service is available.
D. Reduce the banquet's food cost percentage.

6. Which document identifies information employees need to plan special events?

A. Banquet agreement
B. Purchase specification
C. Banquet event order
D. Task breakdown

7. **When is a banquet's customer's guarantee deposit usually paid?**
   A. On the day of the first site visit
   B. On the day the event will be held
   C. During the first meeting with the sponsor
   D. When the number of customers is finalized

8. **What part of a banquet contract relates to the number of meals to be provided?**
   A. Cancellation clause
   B. Indemnification
   C. Purchase order
   D. Guarantee

9. **When is the last payment for a catered event due?**
   A. One week before the event
   B. One month before the event
   C. When the contract is signed
   D. At the conclusion of the event

10. **Which position is typically *not* supervised by the catering site manager?**
   A. Bartender
   B. Food server
   C. Food runner
   D. Kitchen chef

10 Food and Beverage Management: Analysis and Decision Making

INSIDE THIS CHAPTER

- Quality as an Improvement Philosophy
- A Close Look at Financial Analysis
- Corrective Action Process
- Procedures for Implementing Change

CHAPTER LEARNING OBJECTIVES

After completing this chapter, you should be able to:

- Explain how enhanced quality should be the focus of an operation's improvement philosophy.
- Explain a three-step process for analyzing a restaurant or foodservice operation and establishing financial priorities.
- Describe procedures for corrective action plans.
- Describe basic procedures for implementing change.

KEY TERMS

bottom line, p. 285

calendar year, p. 286

continuous quality improvement (CQI), p. 300

fiscal year, p. 286

fixed cost, p. 288

food cost percentage, p. 285

income statement, p. 289

mixed cost, p. 288

net income, p. 293

profit percentage, p. 293

quality system, p. 282

revenue, incremental, p. 287

rollout, p. 301

spot checking, p. 283

supply and demand, law of, p. 287

variable cost, p. 288

CASE STUDY

"Our boss is really making a mistake this time," said Jaren, a server at Indigo Door Cafe.

"You've got that right," replied Stacy. "I can't believe that he thinks the best way to reduce labor cost is to make us wait on more tables. The labor cost will go down because he is paying for fewer hours, but I'll bet that the customer complaints will increase. This plan, like so many others in the past couple of months, will just not work! I wish he had talked to us before he made this decision."

1. How could the manager have used a participative management style in deciding what to do and preparing employees for changes?
2. What alternatives would you consider as a manager if dining-room labor costs were too high?

QUALITY AS AN IMPROVEMENT PHILOSOPHY

A successful operation maintains a high level of quality at all times because it has systems in place to build and maintain quality. There are many components to a good meal: correct ingredients, excellent preparation methods, and customer-friendly service. Consistently maintaining quality is a challenge. Establishments should use ongoing systems to monitor quality rather than reacting to customer-related problems as they arise.

Management Decision Making

Many problems arise in busy operations because employees forget things, take shortcuts, or make other minor mistakes. Many of these mistakes can be corrected before customers are affected. However, if the causes are not identified and resolved, simple mistakes can become big problems. Therefore, a system for regularly reviewing operations is needed, and everyone must be part of it. A **quality system** outlines how quality processes and procedures should be implemented, controlled, and maintained. It involves four steps:

1. Establish standards. Managers establish standards as they develop and implement standardized recipes, purchase specifications, food-preparation sheets, and job breakdowns. These and other quality tools become the benchmarks against which food items and work procedures are judged acceptable or unacceptable.
2. Identify defects. Things do not always go as planned, and defects (differences from established standards) occur. Successful operations manage defects by regularly identifying, correcting, and tracking them. For example, managers monitor food products from receiving through production, and they ensure all standards are followed all the time. They also use effective supervision practices to note and correct employee errors, and they inspect food items before they are served.
3. Track defects. Managers use simple note sheets such as the one shown in *Exhibit 10.1* to recall topics to share with employees during the next pre-shift meeting. They review the daily note sheets on a weekly basis to look for patterns of defects. If they notice recurring problems, they address them.
4. Implement a problem-solving strategy to correct defects. Chapter 7 explained one approach that involves defining the problem, determining its cause and possible solutions, selecting the best alternative, developing and implementing an action plan, and documenting the process for future reference.

Manager's Memo

The best managers know that a problem is not resolved even when its cause has been determined and a solution has been agreed on. When appropriate, employees can be involved in the decision-making process. They should also be updated about action plans. Daily line-up meetings or special meetings are ways to do this.

Managers can do several things during these meetings:

- Review the problem and how its cause was determined.
- Explain what needs to be done to correct the problem.
- Evaluate any additional staff suggestions, and discuss their advantages and disadvantages.
- Describe the recommended solution. Demonstrate the correct procedure, if applicable.
- When possible, put staff in charge of training, maintaining, and implementing the changes.

Exhibit 10.1

WEEKLY REVIEW OF PRODUCT DEVIATIONS

Chef on duty/manager on duty: ______________________ Date: ______________________ Shift: ______________________

Day of week: ______________________

Circumstances to note (weather, special event, etc.) ______________________

By whom	Item noted	Time	Deviation, problem	Resolution	Suggestions
BN	No garnish on Grilled Salmon	6:15 p.m.	Parsley not in dish	Prep cook to get parsley from storage, put in garnish tray	Have more parsley on hand before shift starts
AL	NY Steak not steaming	6:30 p.m.	Finished too early	Refired it	
BN	No garnish on soup of the day	6:40 p.m.	Ran out of garnish (crackers)	Used garnish from special soup	Higher par level for garnish

Evaluating Quality

Products should be evaluated for quality on a regularly scheduled basis. Sometimes product quality issues evolve so slowly that employees and even managers do not recognize them. Customers, on the other hand, will notice if they have not visited the establishment for a while.

Regular review of standardized recipes and food products is one way to maintain product quality. This is called **spot checking**. For example, the manager and chef might decide to learn how well standardized recipes are being followed. They select recipes to test and determine how to administer the spot check during shifts.

THINK ABOUT IT . . .

Many managers use informal spot checking by ordering an item during a break. This allows them to routinely compare items to standards.

Do you think meals for a manager represents the food served to customers?

A spot checking process should involve several strategies:

- Ordering the same meal during two different shifts using the same server
- Evaluating the plating, speed of production and service, and quality (*Exhibit 10.2*)
- Noting how closely the standardized recipe was followed
- Observing the taste and how well the item meets expectations
- Determining whether the meal was the same during both shifts

Exhibit 10.2

A product evaluation sheet such as the one shown in *Exhibit 10.3* can be used to take notes about the meals used for the spot check. It is important to explain the results to the employees, including problems discovered and any changes that will occur as a result of the evaluation.

Exhibit 10.3

ONGOING PRODUCT EVALUATION SHEET

Item ______________________ **Date** ______________

Evaluator ______________________

Directions: Order the same item from at least two different shifts. Complete each cell of the table for these factors to be sure that quality is consistent.

Product Checklist	Manager:		Shift:	Chef:		Shift:
	Yes	No	Comments	Yes	No	Comments
Item delivered properly						
Aroma was attractive						
Appearance—plating was attractive, dish was clean						
Dish looks like menu description						
Garnish was correct						
Temperature was correct						
Seasoning was correct						
Flavoring was correct						
Texture was correct						
Doneness was correct						
Other considerations						
Note if substitutions were made and why						
Were any seasonal factors present?						
Other factors						

What changes need to be made?

____

____

Points to make at preshift meeting

____

____

A CLOSE LOOK AT FINANCIAL ANALYSIS

Managers must be just as concerned about the ongoing financial analysis of their operation as they are about producing and serving menu items of the correct quality. They might ask themselves "What would the financial picture look like if everything were ideal?" They can then compare that ideal benchmark with the actual financial status to assess whether problems exist. Managers can use the three-step process shown in *Exhibit 10.4* to analyze their operation and establish financial management priorities.

Exhibit 10.4

THREE-STEP PROCESS FOR FINANCIAL ANALYSIS

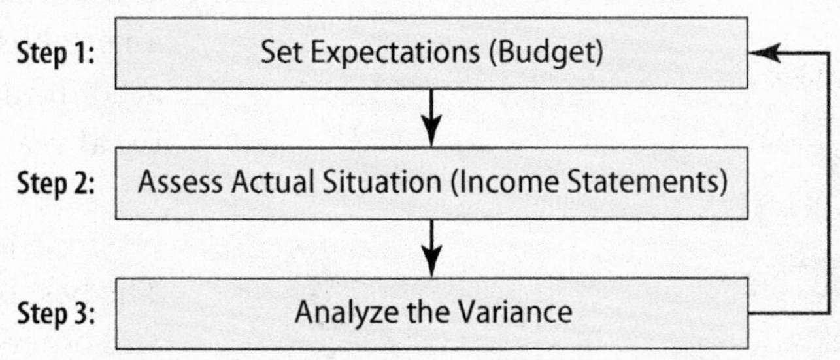

Set Expectations

A student scored 95 on an exam. Is the student pleased or disappointed? This question cannot be answered without knowing how many points the exam was worth. If it was worth 100 points, the score is high, but if the exam was worth 200 points, the score is low.

Unfortunately, some managers attempt to analyze their operation without adequate knowledge. For example, assume an establishment had a 39.6 percent **food cost percentage** last month. This means that 39.6 percent of all food revenue was spent to purchase the food products required to generate that revenue.

Should the manager be delighted or frustrated with these financial results? Just as in the example of student exam results, this question cannot be answered without more information. How much revenue was planned and what food cost percentage was expected?

Restaurant and foodservice managers establish financial goals in their operating budgets. The budget estimates revenues, expenses, and the bottom line. The **bottom line** is a planned profit or another financial goal such as to break even. This goal is expressed on the last line—the bottom line—of the budget and the income statement.

The operating budget is critical to the manager's analysis and improvement responsibilities. The best managers develop and closely monitor budgets. They revise them as business conditions change in efforts to better control financial results.

No manager can predict the future perfectly, and the financial success of an operation is affected by many factors. However, an operating budget is critical:

- It helps managers consider future events and their financial impact.
- It challenges managers to recognize the importance of revenue when projecting expenses.

- It creates a standard against which to compare actual performance.
- It helps managers establish a basic menu pricing structure.
- It communicates a realistic estimate of future financial results to owners so they can evaluate their investment.

ANNUAL OPERATING BUDGET

Most restaurant and foodservice operations create budgets for the 365-day **calendar year** that begins on January 1 and ends on December 31. However, an establishment could budget for a **fiscal year**, which is any 365-day accounting period that does not begin on January 1. For example, the fiscal year might begin on October 1 and end on September 30.

Exhibit 10.5

THE BUDGET TEAM

The best budgets are prepared with input from the general manager and others on the management team (*Exhibit 10.5*). For example, the chef is likely to know about seasonal variations in food costs and the amount of production labor required at projected revenue levels. The dining-room manager can provide insight into service staffing, and the beverage manager can advise about beverage costs based on projected revenues. However, the responsibility to produce the final budget rests with the manager.

BUDGETING REVENUES

Most managers begin the budget development process by estimating revenues, because they impact expenses that will be incurred. Once these are known, profit can be calculated:

Budgeted revenues − Budgeted expenses = Budgeted profit

The budgeted profit level can be met when budgeted revenue levels are achieved and only the amounts budgeted for expenses are spent. If revenues fall short and expenses are not reduced accordingly, profit shortfalls will occur.

If actual revenues exceed forecasted levels, expenses will likely increase. However, if the increases are monitored carefully, increased profits should result. On the other hand, if managers allow expenses to exceed the level required to support the revenue increase, profits higher than those budgeted are less likely.

Revenues are often estimated on a weekly or monthly basis and then combined to create the annual revenue budget. This is an effective strategy because many establishments have seasonal revenue variations. For example, an operation in a shopping mall may generate higher revenues in November and December because of holiday shoppers.

Forecasting revenue is not an exact process, but several basic steps can be used:

Step 1: Review revenue records from previous years.

If an operation has been open for a period of time, its revenue history may help predict future revenue levels.

Step 2: Evaluate internal changes.

Some activities undertaken by the managers or staff can affect revenue projections. Consider, for example, a significant change in the type, quantity, and direction of marketing efforts. Improvements in server training programs can increase revenue because of suggestive selling and reduce labor cost due to increased productivity. Other activities such as renovation may affect seating capacity and times when the operation will be disrupted.

Step 3: Evaluate external changes.

New establishments that open and others that close in the same market can impact revenues, as can road improvements or construction. Economic downturns that reduce consumer spending are another example.

Step 4: Estimate the impact of selling price changes.

Some managers think an increase in menu selling prices automatically results in revenue increases. However, if menu prices increase and sales counts decrease, revenues will decline. This is due to the **law of supply and demand**: economic beliefs about the supply of an item and its price relative to its demand. Generally, as the price of an item increases, the demand for the item decreases (although the reverse can also occur).

Step 5: Estimate effect of menu changes.

Some managers expand or revise menu offerings to increase revenues. Sometimes these efforts are successful, and there are **incremental revenues**: revenue increases beyond current revenues. In other cases, short-term increases are achieved, but longer-term revenues do not increase because the new items are purchased by the same customers who had previously purchased other menu items.

It is not possible to forecast an operation's exact revenues one year in advance when the operating budget is developed. With practice and good information, however, some managers can attain budget forecasts that are a close estimate of actual results.

BUDGETING EXPENSES

After budgeted revenues are forecasted, the next step is to estimate expenses. Some costs, such as rent, remain unchanged regardless of revenues generated. If monthly rent payments are $7,000, the annual rent expense is simple multiplication:

$7,000	×	12	=	$84,000
Rent per month		**Months**		**Annual rent**

Rent is a **fixed cost** because it does not change as revenue changes.

Other expenses increase or decrease as revenue changes. Consider the cost of cloth napkins. As more customers are served, more napkins are used and laundry charges are higher. The napkins are a **variable cost**: a cost that changes in proportion to revenues. Food, beverages, and many other expenses are variable costs.

Still other costs are **mixed costs**: they contain both fixed and variable components. One example is labor. Some employees are paid a salary and receive the same pay if the operation is open or closed, busy or slow. Other employees are paid an hourly wage, and more or fewer waged employees are required depending on the revenue volume.

Budgets should be prepared to reflect all expenses incurred by the operation. Managers recognize the need to budget for food, beverage, and labor expenses, but many other types of expenses can be incurred. Some common expenses are shown in *Exhibit 10.6*.

Exhibit 10.6

COMMON EXPENSES

- Utilities
- Uniforms
- Dry cleaning
- Linens and linen rental
- China
- Glassware
- Silverware
- Kitchen utensils
- Cleaning supplies
- Paper supplies
- Menus and wine lists
- Extermination services
- Flowers and decorations
- Music and entertainment
- Selling and promotion costs
- Sales representative services
- Travel expense for sales calls
- Telephone service
- Postage
- Advertising
- Outdoor signs
- Civic and community donations
- Waste (garbage) removal
- Stationery and business cards
- Data processing costs
- Insurance
- Commissions on credit card charges
- Bank charges
- Repairs and maintenance
- Gardening and grounds maintenance
- Real estate taxes
- Personal property taxes
- Other municipal taxes
- Franchise fees

Assess the Actual Situation

After the manager is aware of the financial expectations stated in the budget, the next step is to determine actual operating results. The results are reported in the establishment's income statement, which summarizes the actual profitability of the operation each month.

The income statement reports the operation's efficiency and profitability. It is the most important financial statement for managers. It allows them to learn if they have achieved the profit levels anticipated and, if not, where improvements are needed. In many companies, managers are evaluated for promotions, pay raises, and increased responsibility based on the income statement.

Income statements, like operating budgets, show revenue and costs in dollar amounts and as a percentage of total revenues. This is a common practice that stems from the view of many managers that revenue and cost percentages are good guides to a manager's effectiveness. For example, a manager might have a 30 percent food cost goal. Computing the food cost percentage monthly would be a good way to monitor the manager's ability to achieve the goal.

There are no "best" income statement goals that can be used by all operations. Food, labor, and other expense percentages may be similar in some industry segments, but the financial information that means the most is the one developed specifically for the establishment. For example, the manager of a fine-dining restaurant that features tableside food preparation and high levels of personal service will spend more money on labor than will a quick-service property. An operation housed in an older building is likely to incur more repair and maintenance costs and higher utility costs than its newly constructed counterpart in the same segment of the industry.

What is consistent among all operations is the profit formula that the income statement details:

Revenues − Expenses = Profits

If profits are too low, the manager must increase revenues, decrease expenses, or both. Information in the income statement can show managers which alternative is most likely to achieve the desired results.

Analyze the Variance

In analyzing the variance, the manager compares the financial plans from the approved operating budget with the actual operating results in the income statement. This should be done on a regular basis to determine where any variances exist. A potential financial problem exists when there is a significant variance between operating budget and income statement information. Then the manager must determine whether further analysis is necessary.

In chapter 7, the difference between budgeted and actual results was defined as a variance. It may be expressed in either dollar or percentage terms and can be either positive (favorable to the operation) or negative (unfavorable to the operation). A positive variance occurs when the variance is an improvement on the budget, and revenue is higher or expenses are lower. A negative variance occurs when actual results are less than budget expectations, and revenue is lower or expenses are higher.

For example, if the budget for landscape services is $1,000 for a given month but the actual expenditure is $1,150, the variance may be expressed as a dollar amount of $150, or as a percentage of the original budget:

($1,150	−	$1,000)	÷	$1,000	=	15%
Actual expense		**Budgeted expense**		**Budgeted expense**		**Variance percentage**

Since the variance is negative, some managers would express it as (15 percent) or −15 percent.

The manager must identify the *significant* variances between planned and actual operating results. These are the variances that represent a large difference in dollars or percentage and therefore warrant further investigation.

Significant variance is an important concept because not all variances should be investigated. For example, a manager has budgeted a $3,000 December utility bill, and it actually totals $3,025. Given the difference of only $25 and the difficulty of estimating utility expenses, this probably does not represent a significant variance.

Alternatively, if the manager had estimated office supply usage at $100 but the actual cost was $500, this represents a much more significant difference. It is also an expense that can probably be better controlled, so an investigation may be in order.

Managers must decide what represents a significant variance based on their knowledge of the operation. Small percentage differences can be important if they represent large dollar amounts. Similarly, small dollar amounts can be significant if they represent large percentage differences from planned results.

Exhibit 10.7 shows a comparison of income statement information (columns 2–3) and budget estimates (columns 4–5) for the month of January. The categories of revenues and expenses and the order are the same in each document to make the comparison easier.

It is important to analyze actual and budgeted results in each of the income statement's three major sections. The statement is divided into revenues, expenses, and profits.

Exhibit 10.7

INCOME STATEMENT AND BUDGET FORECAST COMPARISON

	January			
(1)	(2) Actual	(3) %	(4) Budget	(5) %
REVENUES				
Food Revenue	$ 88,000	69.8%	$ 98,000	68.5%
Beverage Revenue	38,000	30.2	45,000	31.5
Total Revenues	**$126,000**	**100.0%**	**$143,000**	**100.0%**
COST OF SALES				
Food	$ 30,850	35.1%	$ 33,124	33.8%
Beverages	11,000	28.9	11,250	25.0
Total Cost of Sales	**$ 41,850**	**33.2%**	**$ 44,374**	**31.0%**
Gross Profit	**$ 84,150**	**66.8%**	**$ 98,626**	**69.0**
OPERATING EXPENSES				
Salaries and Wages	$ 39,180	31.1%	$ 41,613	29.1%
Benefit Allocation	9,200	7.3	10,439	7.3
Marketing	4,670	3.7	5,720	4.0
Utilities	2,900	2.3	3,289	2.3
Administrative Expense	4,150	3.3	3,289	2.3
Maintenance and Repair	2,710	2.2	2,860	2.0
Taxes/Insurance	2,810	2.2	2,810	2.0
Depreciation	6,210	4.9	6,210	4.3
Total Operating Expense	**$ 71,830**	**57.0%**	**$ 76,230**	**53.3%**
Operating Income	**$ 12,320**	**9.8%**	**$ 22,396**	**15.7%**
Interest	$ 2,600	2.1%	$ 2,600	1.8%
Income before Income Taxes	**9,720**	**7.7**	**19,796**	**13.9**
Income Taxes	4,210	3.3	8,100	5.7
NET INCOME	**$ 5,510**	**4.4%**	**$ 11,696**	**8.2%**

REVENUE ANALYSIS

Revenues are the first area to examine when comparing actual to budgeted results. If revenue is significantly below projected levels, profit goals are likely to suffer. Also, when revenues vary from projections, variable costs will change. They will be less than budgeted if revenues are less. In contrast, when revenues fall short of budgeted levels, fixed costs such as rent and mixed costs such as labor will likely represent a larger-than-budgeted percentage of total revenue.

Another reason to examine revenues is that understanding how actual revenue has varied from budget estimates will help managers better understand actual expenses and profits. For example, if revenue exceeds forecasts, increases in operating expenses (a variable expense) relate at least partly to increased revenue levels. The impacts on budgeted expenses that result from variations in revenue are listed in *Exhibit 10.8*.

Exhibit 10.8

REVENUE VARIATION AND FIXED AND VARIABLE EXPENSES

Budget Impact	If Actual Revenue Greater Than Budgeted	If Actual Revenue Less Than Budgeted
Fixed Expenses		
In dollars	None	None
As a percentage of revenue	Decreases	Increases
Variable Expenses		
In dollars	Increases	Decreases
As a percentage of revenue	May fluctuate*	May fluctuate*

*100 percent variable expenses increase or decrease in direct proportion to revenue increases or decreases. However, most expenses have at least some fixed component. Therefore, as revenue increases, variable expense percentages may actually decline. Similarly, when revenue fails to meet budget, some increases in the variable expense percentages will likely result.

Looking back at *Exhibit 10.7*, note that the operation has had a shortfall in both food and beverage revenue. When revenue consistently falls short of forecasts, managers must use marketing strategies to reverse this problem. Strategies can include the use of coupons, increased advertising and marketing efforts, price discounting, and the promotion of specials.

EXPENSE ANALYSIS

Identifying significant expense variances is a critical part of the budget monitoring process. While managers may be limited in their ability to influence revenue, many expenses are under their direct control. Some variation between budgeted and actual expenses can be expected because, as noted, some operating expenses vary with revenue volumes. Variances can, however, tell much about operational efficiencies. Experienced managers know that a key to ensuring profitability is properly controlling operating costs.

Costs in income statements and operating budgets are typically expressed in dollars and as a percentage of revenue:

Expense ÷ Total revenue = Expense percentage

As seen in *Exhibit 10.7*, each operating expense such as salaries and wages is calculated this way. However, the food and beverage percentages are calculated differently:

Cost of sales: Food ÷ Food revenue = Food cost percentage

A similar calculation would be used to determine the expense for "Cost of sales: Beverage." Food and beverage calculations use the revenues that each generated because these cost ratios are important to total profitability. In *Exhibit 10.7*, food costs were higher than budgeted at 35.1 percent actual versus 33.8 percent budgeted. Beverage costs were also higher at 28.9 percent actual versus 25.0 percent budgeted.

Note that while total operating expenses were budgeted at 53.3 percent of total revenue, the actual results were higher at 57.0 percent. Specific operating expenses must be evaluated and significant variances analyzed using the budget monitoring process introduced in this chapter. Note that most operating expenses increased over budget estimates. This issue should be examined closely.

One area of concern is that actual marketing costs in dollars and percentages were lower than budgeted. Higher marketing expenditures might have improved revenues. Note also that two fixed costs (depreciation and interest) did not vary in dollars but produced a higher cost percentage because the fixed dollars were spread over a smaller revenue base.

PROFIT ANALYSIS

When the actual revenues and expenses on the income statement differ from those on the operating budget, the result is often a change in the net income on the income statement. **Net income** represents the profit made in a business. In restaurant and foodservice establishments, profits are also typically stated as a **profit percentage**, which is the profit or loss expressed as a percentage of total revenue.

As seen in *Exhibit 10.7*, the net income for the month was $5,510 or 4.4 percent of total revenues. This is less than the $11,696 (8.2 percent of revenues) that was budgeted. This relates directly to the lower revenues and, in some categories, higher expenses. The operation's inability to meet budget goals means that the budget was poorly developed, internal or external conditions changed, or the managers were ineffective. Regardless of the causes, corrective action plans are needed to prevent even more serious problems.

Profits can be expressed in percentages and dollars. Which is best? There is a saying that managers bank dollars, not percentages. Evaluating profit dollars is important, but care must be taken. Why? The operation analyzed in *Exhibit 10.7* achieved a net income profit percentage that was lower in both dollars and percentage than the budget forecast. Consider a scenario in which the revenue remained the same, but the actual net income was 8.3 percent. This net income percentage would be higher than the 8.2 percent budgeted. However, the total dollars of profit ($126,000 actual revenues × 8.3% = $10,458) would still be less than the budgeted profit of $11,696. There would be a lower dollar amount and a higher profit percentage.

Now consider two establishments with different levels of revenue. In the first, the manager achieves $1,000 in profits on $10,000 of revenues:

$1,000 ÷ $10,000 = 10% of revenue = Profit

In the second, the manager achieves profits of $1,100 on revenues of $50,000:

$1,100 ÷ $50,000 = 2.2% of revenue = Profit

The second manager achieved a greater dollar level of profits, but the first achieved a higher profit percentage. It is likely that the first manager operates an establishment with lower food, labor, and other costs. The second establishment has a higher dollar amount of profit and a lower profit percentage. Since the manager puts profit dollars not profit percentages into the bank, the second manager operates a more profitable operation.

These examples show that neither dollars nor profit percentage tell the whole story. Effective managers review both profit percentages and dollars when comparing their operating budget with their income statement.

Budget Revisions

Managers should consider their budgets active documents that may be subject to change. Budgets should be regularly reviewed and modified as better information replaces the information available when the original budget was developed. This is especially true when the new data significantly affects revenue and expense assumptions.

The budget should be reviewed any time managers believe the assumptions it is based on are unfounded. Consider an establishment's owner employing 50 full-time staff members. Each staff member is covered under a group health insurance policy. Last year, the owner, who pays 50 percent of the insurance cost, paid $200 per month for each employee:

50	×	$200	=	$10,000
Employees		**Per employee**		**Total monthly cost**

When the budget was developed, the manager assumed a 10 percent increase in health insurance premiums. If the premiums actually rose 20 percent, employee benefit costs will be greater than projections each month. The owner faces several choices:

- Modify the budget to reflect the change in premiums.
- Change the amount contributed to stay within the budget.
- Change (reduce) benefits to stay within the costs allocated.

Regardless of the owner's decision, the budget, if affected, should be modified. *Exhibit 10.9* shows some situations that generally require a reexamination of the budget with changes made as necessary.

Exhibit 10.9

FACTORS THAT MAY REQUIRE BUDGET REVISIONS

- The opening or closing of a major direct competitor
- A significant long-term or permanent increase or decrease in the price of major menu ingredients
- Significant and unanticipated increases in fixed expenses such as utilities, insurance, or taxes
- A management change that significantly alters the skill level of the management team
- Unplanned road construction that significantly affects customers' ability to reach the establishment
- Natural disasters such as floods, hurricanes, or severe weather that significantly affect forecasted revenues
- Significant changes in the establishment's operating hours
- Permanent changes in service style that appreciably affect labor costs
- Changes in financial statement formats or bases for allocation of financial resources
- Significant changes in variable costs that offset revenues
- The loss of especially skilled employees

CORRECTIVE ACTION PROCESS

After significant variances between planned and expected operating results are identified, they must be managed. Managers can use a corrective action plan that involves the same basic steps described in the problem-solving model in chapter 7. The steps are useful to address employee performance management (chapter 7) and financial analysis because they are a model for the decision-making process.

Identify Problems

Effective managers recognize that it is their responsibility to address problems that hinder the attainment of financial goals. This fact may seem obvious, but some managers think that they are already doing their best and it is not possible to improve.

Others, especially those with extensive experience, may believe there is a shortcut to being successful such as "just take care of the customers" or "focus on food costs." This may cause them to pay attention to only part of the operation. Today's managers recognize that they must both please customers and reduce the costs necessary to do so.

Managers are busy, and they cannot do everything they would like to do. Therefore, they must identify the problems and then address the most important concerns first. Careful analysis of budget plans and actual operating results expressed in the income statement is an excellent way to identify financial problems.

If managers identify more than one financial challenge, which should be addressed fist? In other words, which is more important: controlling revenue or controlling costs? Addressing food costs, labor costs, or another type of cost? Managers must make decisions and set priorities. *Exhibit 10.10* gives some suggestions.

As seen previously, a 2 percent variance in food costs represents $7,000 in higher-than-expected costs and lower profits. By contrast, a 5 percent variance in beverage cost results in $5,000 in higher costs. Clearly, after any "quick fix" with the beverage operation, the manager's attention must be directed to food control activities.

To summarize the points in *Exhibit 10.10*, a revenue increase is offset by the variable costs required to generate the revenue. By contrast, a cost decrease falls completely to the "bottom line." *Exhibit 10.10* emphasizes that it is not the percentage of variance that is most important in determining control activities. Instead, it is the number of dollars the variance represents.

OPEN FOR BUSINESS

MANAGER'S MATH

Consider the following situation for the month of March:

Revenue	$14,500
Variable costs	75%
Fixed costs	18%

How much profit would be generated if variable costs were reduced by $850?

(Answer:

Revenue	*$14,500*
Variable costs	*(10,025)*
(75% × $14,500 − $850)	
Fixed costs	*(2,610)*
(18% × $14,500)	
Profit	*$1,865)*

Exhibit 10.10

ADDRESS THE BIGGEST PROBLEMS FIRST

A. Is it preferable to increase revenue or decrease variable costs? Consider the following example.

Assume	Current Data	Increase Revenue by $2,000	Decrease Costs by $2,000
Revenues	$10,000	$12,000	$10,000
Variable costs (70%)	(7,000)	(8,400)	(5,000)
Fixed costs: current (20%)	(2,000)	(2,000)	(2,000)
Profit	$ 1,000	$ 1,600	$ 3,000

Increasing revenue by $2,000 yields only a $600 increase in profit ($1,600 – $1,000); decreasing variable costs by $2,000 increases profit by $2,000 ($3,000 – $1,000).

B. Which of the following costs should be managed first? Which is the priority?

	Actual	Budget	Difference
Food cost	34%	32%	(2%)
Beverage cost	30%	25%	(5%)

At first it appears that beverage cost represents the bigger problem: a 5% variance compared to only 2% for food cost. However, more details may affect that opinion:

Food			
	Actual	Budget	Difference
Food revenue	$350,000	$350,000	
Food cost %	34%	32%	
Food cost	$119,000	$112,000	($7,000)

Beverage			
	Actual	Budget	Difference
Beverage revenue	$100,000	$100,000	
Beverage cost %	30%	25%	
Beverage cost	$ 30,000	$25,000	($5,000)

Make Decisions

After the manager has determined a problem to address, a decision-making process can be used. *Exhibit 10.11* shows how managers can reduce problems identified in the financial analysis process.

Exhibit 10.11

EXAMPLE OF DECISION-MAKING PROCESS

Step	Example
Step 1: Define the problem	Check average for food has been declining for the last three months.
Step 2: Determine the cause	Note the following: • Suggestive selling • Menu problems including pricing • Errors in procedures for calculating customer check average • Server theft
Step 3: Determine and evaluate alternative solutions	• Customer shoppers did not respond to suggestive selling; need training. • Menu recently redesigned; customer counts are up slightly. • An auditor has found no bookkeeping problems suggesting error or employee theft.
Step 4: Select the best alternative	Implement a suggestive selling program.
Step 5: Develop and implement an action plan	Train service staff and implement a contest: All servers with a specified minimum customer check average win a prize.
Step 6: Evaluate and document the solution	Determine how much the customer check average increases after the suggestive selling is rolled out.

Some basic principles are used in the steps shown in *Exhibit 10.11*:

Step 1: Define the problem.

Sometimes problems such as lower revenue levels and increasing costs are obvious. At other times it is more difficult to identify the problem. Consider the view of some managers that an increasing number of job applicants enter the workforce with a lower work ethic than earlier generations. There are probably societal, cultural, and other components of this issue that make it hard to specifically identify the problem.

To help define the problem, some managers ask themselves “What would the situation be like in the absence of the problem?” Their answer allows them to think about the problem in a different way.

Unfortunately, problems do not occur one at a time. Instead, managers are often faced with several problems at the same time. Then they must set a priority.

Step 2: Determine the cause.

What can be done to address the problem? The manager should have some answers, but the employees may as well. The use of cross-functional teams can help identify potential solutions. For example, a problem about customer complaints regarding slow service could be addressed by the manager, servers and cooks, and perhaps others.

Step 3: Determine and evaluate alternative solutions.

Factors including costs, ease of implementation, and impact on other work processes can help evaluate the alternatives.

Step 4: Select the best alternative.

Often, the best solution involves aspects of several possible alternatives.

Step 5: Develop and implement an action plan.

Employee training (*Exhibit 10.12*), purchase of necessary equipment or tools, and changes in work procedures are among the strategies that may be useful in implementing the chosen alternative.

Exhibit 10.12

Step 6: Evaluate and document the solution.

If the manager has already considered what the situation would be like if the problem no longer existed back in step 1, this becomes easier. Sometimes solutions help but do not totally resolve the problem. In other words, as a result of the decision-making process, the problem's impact is reduced. If the problem is still significant relative to others, the manager may repeat the decision-making process.

PROCEDURES FOR IMPLEMENTING CHANGE

Perhaps the analysis and corrective action processes will determine that an existing procedure must be consistently used. Perhaps, however, the manager will recognize that new procedures will be necessary to keep revenue and costs in line with budget estimates. At this point the manager must become a master of implementing change.

One of the first obstacles that potentially must be overcome is that of affected employees resisting changes. Employees who are used to doing something one way may resist changes in standard operating procedures. They may be uncertain about how they will be affected by the change, or they may not want to take time to learn new ways of doing things. They may also be concerned about the closer coaching and supervision interactions necessary as the change is implemented and evaluated. If this natural resistance is understood and addressed, the manager is more likely to be successful in the change effort.

Overcoming Resistance to Change

There are a number of strategies a manager can use to overcome resistance to change. All will be more effective if the manager has a history of involving employees and explaining, defending, and justifying why a change is necessary. It is also helpful if the manager has historically been right: The situation has usually been better after the change than before it was made.

There are several strategies for reducing resistance to change:

- Involve employees in the decision-making process. A participative management style improves the process and makes implementation easier because the decision involved the employees. In other words, the employees have a buy-in; they have invested their creative ideas in the solution and want to prove that their solution will be a good one.
- Inform employees in advance about changes that will impact them.
- Select an appropriate time to implement the change. Trying something new during an extremely busy shift is rarely a good idea.
- Share past successes; review related changes that have benefited the employees and the organization.
- Reward employees for sharing ideas that benefit the operation and the employees.

Managing Change

Managers who believe in the quality management process will be leading their staff members through a plan of **continuous quality improvement (CQI)**. This involves ongoing efforts to better meet or exceed customers' expectations, and to define ways to perform work with better, less costly, and faster methods.

These managers recognize that regardless of how small a change may be, the operation is improved by any change that helps it better meet its mission and goals. Employees working in a CQI environment will be conditioned for change, look forward to it because they know its benefits, and be active participants in it. CQI typically works "from the bottom up" because the employees closest to the situation are likely to have improvement suggestions. Their ideas about increasing revenues and controlling costs, for example, can be very creative. In fact, the manager really becomes a facilitator. After defining a problem, he or she uses a team approach to analyze, take corrective action, and implement necessary changes.

Quality Improvement in Action: Implement Rollouts

Managers and their teams who implement corrective action plans often find themselves revising policies and operating procedures. They might also add, remove, or revise menu items. The term **rollout** describes the process of introducing something new in an operation.

Effective planning for a rollout is very important. If it is not properly executed, the results can impact costs and revenues, both of which could result in lower profits. Also, customer satisfaction and employee performance may be affected. Customers may be disappointed and never return, and employees may have less confidence in the managers to make good decisions,

Several basic procedures are useful for implementing rollouts:

Step 1: Identify rollout changes.

The analysis and corrective action processes discussed in this chapter should help identify what changes are necessary.

Step 2: Plan rollout details.

A well-planned rollout addresses timelines, training needs, new equipment and tools, revised external and internal marketing plans, and other specifics.

Step 3: Consider staff training if necessary.

Training is almost always necessary in the rollout process. Training materials may need to be developed or purchased, and the requirements addressed in chapter 7 become important. The objective of the training should address what employees must know or be able to do when the change is made.

Step 4: Implement rollout.

The rollout process must consider the timeline for every detail in step 2. It must also identify the persons responsible for specific activities in the rollout plan.

Step 5: Collect rollout data.

This step will be easier if the manager and team considered how to determine success in steps 1 or 2. If, for example, a new procedure involves suggestive selling to increase the customer check average (*Exhibit 10.13*), the check average should increase as a result of the strategies implemented. If a new menu item is introduced, data should include customer comments, sales history information, and comments from production and service employees.

Exhibit 10.13

Step 6: Evaluate rollout data.

If the objective of the change is measurable, the rollout data (step 5) can help managers determine if the rollout created the desired results.

Step 7: Adjust rollout if needed.

The extent to which adjustments will be necessary depends on how well rollout objectives were attained (step 6).

Step 8: Record rollout results.

If the rollout is successful, this step will be very important. For example, training programs for new employees must reflect revised procedures. If menu items are involved, standardized recipes, purchase specifications, and quality standards used to evaluate products will require revision.

This process will be applied in a similar manner for different types of rollouts. *Exhibit 10.14* provides an example of the steps involved in the rollout of a new

Exhibit 10.14

EXAMPLE OF ROLLOUT: NEW PROCEDURE

Consider how this plan can be used for the rollout of the suggestive selling program in *Exhibit 10.11*:

Step 1: Identify rollout changes.

A suggestive selling program will be implemented to increase the customer check average.

Step 2: Plan rollout details.

The plan is to increase the customer check average by 50 cents per customer within six weeks after a training program is completed. A training program developed by the headquarters of the organization is available. The dining-room manager will conduct the program and will be given one week to study the details and produce any necessary training materials. Times for all dining-room employees to attend the program will be scheduled. A contest is planned that allows all servers who achieve a customer check average of $16 during one week to receive movie passes from a local theater.

Step 3: Consider staff training if necessary.

The training program will be one hour long and will be taught to different groups of servers over a two-day period.

Step 4: Implement rollout.

A specific evening shift will be designated as the start period for the suggestive selling program.

Step 5: Collect rollout data.

The operation's point-of-sale (POS) equipment will be used to generate customer check average data for each server.

Step 6: Evaluate rollout data.

The establishment manager and dining-room manager will review and calculate customer check average information for each server each week.

Step 7: Adjust rollout if needed.

Approximately 70% of the servers achieved the check average goal the first week. The program was considered a success, and the managers will coach employees who did not achieve the goal during the coming weeks.

Step 8: Record rollout results.

Since the suggestive selling program is successful, it will be incorporated into the job description and training materials for servers.

procedure, while *Exhibit 10.15* illustrates what might be involved in a new product rollout.

Managers Are Decision Makers

Managers of restaurants and foodservice operations must make many important decisions as a routine part of their jobs, and their performance is impacted by the quality of those decisions. The chances of these important decisions being good ones improve when analysis and planning are done to help ensure that the implementation process will be successful. Also, analysis and evaluation after the implementation will be helpful to guide managers as future decisions are made.

If, for example, the desired result is achieved, this decision might be repeated in a similar situation. Likewise, if the decision did not yield the desired results, the manager will know what did not work. Then he or she can consider other alternatives when the same or a similar situation arises.

Exhibit 10.15

EXAMPLE OF ROLLOUT: NEW MENU ITEM

The rollout procedures discussed in this section can also be used when a new menu item is introduced:

Step 1: Identify rollout changes.

The decision has been made to eliminate one menu item and to replace it with another. The new item must have a higher contribution margin and be prepared to order to eliminate expensive leftovers.

Step 2: Plan rollout details.

The new item is expected to generate at least 5% of all entrée sales. The chef will collect and test possible recipes over the next three weeks. The purchaser will identify quality concerns and begin discussions with potential vendors during the next two weeks. At the same time, production staff will be trained and allowed to practice production of the menu item. Servers will sample the test results to obtain product knowledge about the new item.

Step 3: Consider staff training if necessary.

Production employees will review the final recipe. Cooks who will be preparing the item will be given follow-up training.

Step 4: Implement rollout.

The new menu item will be featured for two weeks as a "Chef's Special" that is not on the formal menu.

Step 5: Collect rollout data.

POS system information will be collected for the new menu item. Data will include sales counts and estimated contribution margins.

Step 6: Evaluate rollout data.

The POS and contribution margin information will be studied to determine if the sales requirement (5% of dinner entrées) has been met.

Step 7: Adjust rollout if needed.

The rollout results are better than expected, so internal selling tools are developed including table tents and a menu clip-on. Special information is posted on the operation's Web site.

Step 8: Record rollout results.

The standardized recipe for the new menu item will be added to the operation's permanent recipe file. In addition, the formal menu will be revised to feature this item.

SUMMARY

1. **Explain how enhanced quality should be the focus of an operation's improvement philosophy.**

   A quality system involves several steps: establish standards, identify defects, track defects, and implement a problem-solving strategy to correct defects that occur.

   Spot checking strategies by the manager and chef are useful for evaluating quality. They might, for example, order the same meal during two different shifts and see whether the meal is the same. They should evaluate the plating, speed of delivery, and quality, and confirm that a standardized recipe was followed. Managers may take daily or weekly notes to record their findings and communicate issues to the appropriate staff members.

2. **Explain a three-step process for analyzing a restaurant or foodservice operation and establishing financial priorities.**

   A three-step analysis process for establishing corrective action priorities includes developing a budget, assessing the situation using the income statement, and analyzing for variances.

   Budgets are prepared with input from the management team. The basic process includes estimating revenues and then forecasting expenses (including fixed, variable, and mixed costs) to determine profits. Experienced managers use historical information and their knowledge about internal and external changes to develop budget benchmarks.

   Income statements reflect the operation's efficiency and profitability. Significant variances between budget expectations and actual performance reflected by the income statement require analysis. Analysis should include both dollar figures and profit percentages. If basic assumptions have changed, the budget should be revised.

3. **Describe procedures for corrective action plans.**

   Sometimes corrective action is necessary to reduce variances. First, problems must be identified to determine those that most affect profitability. Managers must then make decisions about how to address unacceptable variances most effectively. One decision-making process involves defining the problem, determining and evaluating solution alternatives, selecting and implementing the best alternative, and evaluating its effectiveness.

4. **Describe basic procedures for implementing change.**

   Employees may resist the changes required. Managers can use several strategies to overcome resistance to change. First, the process will be easier if the manager has involved employees and explained why changes are necessary. Also, change is easier if situations have historically been better after changes than before.

   A continuous quality improvement (CQI) initiative can drive the culture in which change is implemented. Employees will know that regardless of how small a change might be, the operation can better meet its mission and goals.

Several basic procedures can be used to implement rollouts of new or revised policies, procedures, and menu items. This process involves identifying rollout changes, planning rollout details, considering staff training if necessary, and implementing the rollout. Then, rollout data must be collected and evaluated, and the rollout may need adjustments depending on how well its objectives were attained. A final step is recording the rollout results.

APPLICATION EXERCISE

A trendy Italian restaurant features an appetizer pizza that has been popular for several years. The standardized recipe calls for the pizza to be served cold after being cut into 12 equal-sized pieces. The pizza is topped with cream cheese and smoked salmon. The garnish is a dollop of crème fraîche with caviar on all 12 pieces.

Lately, the pizza is being served on hot crust, which makes the cream cheese soft and causes the crème fraîche to melt into the smoked salmon. The presentation is not like the menu description. The temperature and texture are incorrect as well. Some servers have mentioned this to the chef and have stopped recommending the once-popular appetizer. Additionally, the manager has noticed that sales are down for this item.

Write a detailed description of what persons in each of the following positions should do to identify the cause of the problem and correct it: server, cook, and manager. As you develop each employee's role in the problem-solving process, think about three things:

1. What can be done to best ensure improved and consistent quality?
2. What corrective actions should be taken?
3. How should the change be implemented to best ensure that it will be effective and long lasting?

REVIEW YOUR LEARNING

Select the best answer for each question.

1. **How frequently should note sheets that track quality defects be reviewed?**
   A. Hourly
   B. Daily
   C. Weekly
   D. Monthly
2. **What is the process used for regularly reviewing standardized recipes to maintain quality?**
   A. Corrective action
   B. Variance analysis
   C. Spot checking
   D. Rollout evaluation
3. **What financial document is used to assess actual revenues and expenses?**
   A. Operating budget
   B. Cash review
   C. Income statement
   D. Accounting audit
4. **What is the term for a 365-day accounting period that does not begin on January 1 and end on December 31?**
   A. Fiscal year
   B. Income year
   C. Calendar year
   D. Accounting year

5. **Budgeted revenue minus budgeted expenses equals budgeted**
   A. variable costs.
   B. mixed costs.
   C. investment.
   D. profit.

6. **What is an example of a fixed cost?**
   A. Rent
   B. Utilities
   C. Beverage
   D. Labor

7. **What is an example of a variable cost?**
   A. Depreciation
   B. Food
   C. Interest
   D. Insurance

8. **The budgeted food cost is $10,000 and the actual food cost is $12,000. What is the food cost percentage variance?**
   A. 2%
   B. 12%
   C. 15%
   D. 20%

9. **If actual revenue is greater than budgeted revenue, a fixed expense stated in dollars will**
   A. increase.
   B. decrease.
   C. stay the same.
   D. vary with revenue.

10. **What is the best strategy to increase profits without reducing quality of products and services?**
    A. Increase revenue
    B. Increase customers
    C. Decrease investments
    D. Decrease costs

FIELD PROJECT

RESTAURANT AND FOODSERVICE MANAGEMENT IN THE REAL WORLD

In this field project, you will assume the role of the general manager for an Italian-themed establishment. You are planning it with another person who is investing most of the money in the project. Your contribution in the planning and operation will be to use your knowledge and experience that includes, for the last three years, serving as the manager of a high-volume operation.

The investor has purchased a fully furnished 7,500-square-foot building that had been a steakhouse but has been closed for about two years. Good news: It is located in a "restaurant row" in a trendy suburban area. Locals frequently dine out there, and tourists often visit the area in response to the advertising of many properties in the area.

Your experience has taught you many things to implement in the new property, and it has also provided some lessons about other things you don't want to do. The investor is counting on you to help ensure the operation's success. You and she know that many establishments go out of business. One reason is their failure to have basic restaurant and foodservice management principles in place on the day the property's doors open to customers.

PART I: The Menu

Instructions

Visit an Italian-themed, table-service establishment in your area and ask the manager if you can borrow a dinner menu for a class project.

Pretend that the menu you have borrowed is a first or second draft of a menu for your new operation that you and a consulting chef have developed.

Evaluate that menu using the comments from your field project exercise for chapter 1. You can include them on the following form:

EVALUATION OF SAMPLE MENU	
Evaluation Factors (from chapter 1)	**Your Assessment of Sample Menu**
From the establishment manager ____________________ ____________________ ____________________	
From a friend or relative ____________________ ____________________ ____________________	
Things to avoid (personal opinion) ____________________ ____________________ ____________________	
Things to include (personal opinion) ____________________ ____________________ ____________________	

PART II: Standardized Recipes

Calculate each entrée's per-serving cost for the standardized recipe you developed in the field project exercise for chapter 2.

Use current ingredient costs for your area and calculate the per-serving costs. Adjust the recipe to yield 20 servings (same serving size) if you do not select a recipe yielding 20 portions. Hint: *You may find current costs by checking local grocery stores, asking a manager at your college foodservice, or contacting a local supplier for some hard-to-locate prices.*

Name of menu item: ______________________

Ingredient	Amount	Purchase Unit	Cost per Purchase Unit	No. of Purchase Units	Ingredient Costs

Total ______________

Cost per serving = Total cost ÷ 20 = $______________

PART III: Quality Food Production Standards

Review the notes you took as you interviewed an operation manager or chef in the field exercise for chapter 5.

Also, obtain a copy of the food inspection report used by the health inspectors in your county. A suggestion is to use your favorite search engine, enter "(name of your county) public health department," and then type "restaurant inspections" in the site's search box.

1. Use your notes and the inspection report to develop a list of factors that can be included in the manager's or chef's ongoing review of foodhandling procedures and conditions at the establishment.

2. How might the checklist also be used for training and coaching of foodhandling employees?

____

____

____

____

____

____

PART IV: Quality Beverage Management Standards

What are five popular Italian wines that you might offer your customers that complement your menu? Please include red and white wines in your list. Hint: *Review your notes from the field project exercise in chapter 6 to learn ideas about popular Italian wines. You can also enter "Italian wines" in your favorite search engine.*

For each wine, provide information that you can use to train your service employees. Your notes from the field project exercise for chapter 6 can also help you with training ideas. Your examples of training information can include pronunciation, food and wine pairings, and a brief description of each wine's taste.

Wine 1: ____

____

____

Wine 2: ____

____

____

Wine 3: ____

____

____

Wine 4: ____

____

____

Wine 5: ____

____

____

Part V: Facilitating Performance—Job Descriptions

Review the major job tasks for cooks, bartenders, and food servers that you developed in the field project exercise for chapter 7.

1. What are some procedures you would use to train employees to perform some of the tasks you discovered in your field project exercise for Chapter 7? Hint: *If possible, interview the manager of an establishment that opened recently in your community.*

   A. Procedures to train cooks:

   B. Procedures to train bartenders:

   C. Procedures to train food servers:

PART VI: Two Final Questions

1. Now that you have read through this ManageFirst Book and worked through the exercises in this field project, do you think it is more challenging to operate a successful establishment than you did before you learned about some of the principles of food and beverage management? Why or why not?

2. Many operations that open do not generate the profit levels hoped for by their owners and managers. Still others fail and go out of business. What are the five highest-priority suggestions you would make to those wanting to open a restaurant or foodservice operation to help them be successful?

Suggestion 1: ______________________________

Suggestion 2: ______________________________

Suggestion 3: ______________________________

Suggestion 4: ______________________________

Suggestion 5: ______________________________

GLOSSARY

Accompaniment An item such as salad, potato, and other choices included as part of the entrée selling price.

Action plan A series of steps that will be taken to resolve a problem.

Aging One of two methods of tenderizing beef: Wet aging involves placing small portions of beef in a plastic bag without air, sealing the bag, and refrigerating for a specific period of time. Dry aging is done in special coolers in which air is circulated at precise temperatures for the time determined by package size and other factors.

"A" item A very expensive item.

Àla carte menu A menu that has different prices for each menu item.

Anaphylaxis A potentially life-threatening allergic reaction that can cause a drop in blood pressure, loss of consciousness, and even death.

As purchased (AP) The weight of a food item before it is cooked.

Backorder An order that a vendor cannot fill or ship immediately.

Bag-in-box (soft drink syrup container) A 5-gallon syrup container in which the soft drink syrup is sealed in a plastic bag that is then placed in a cardboard box for easy transporting.

Banquet An event in which a sponsor pays an establishment to provide specified food and beverage services to all event attendees.

Banquet agreement A contract between the operation and the event's sponsor, which specifies the responsibilities of both parties.

Banquet event order (BEO) A document used by sales, production, and service staff to detail all banquet requirements.

Batch cooking The production of items in small volumes (batches) to maximize food quality by reducing holding times until service.

Behind-bar par level The number of bottles of each item that should be behind the bar at any time.

Benefits Indirect financial compensation paid to attract and retain employees.

Bin number (wine) A number that tells the location in a wine storage area where a specific wine is stored.

Blind tasting A tasting during which the raters do not know the recipe's ingredients or preparation methods when they sample the items.

Blood alcohol content (BAC) The amount of alcohol absorbed into a person's bloodstream.

Bona fide occupational qualifications (BFOQs) The realistic range of skills or credentials needed to perform the essential job functions.

Bottom line A planned profit or another financial goal such as to break even.

Brand The combination of qualities that makes an establishment's products and services different from those of its competitors.

Broad line vendor A vendor that sells a wide range of products, often with few alternatives in each category.

Broken case storage An area used to store opened purchase units of products that have been issued.

Buffet A style of foodservice in which customers select the menu items and portion sizes they prefer as they pass along one or more serving counters.

Calendar year A year that begins on January 1 and ends on December 31.

Calibrate To check or verify, such as oven temperature.

Call brand (spirits) A specific brand of liquor requested by a customer.

Cancellation clause A document that indicates any financial penalties for cancellation at certain times before the banquet date.

Capital The amount of an owner's money invested in a business that could be used for other purposes.

Cash bar An event during which alcoholic beverages are purchased by the event attendees.

Cash flow The amount of money needed to pay bills when due.

Catering The production and service of food and beverages at a location other than the establishment responsible for the event.

Cellar temperature A constant storage temperature between 55°F and 60°F (13°C to 16°C).

Centralized purchasing A system in which purchasing requests are routed to those with specialized responsibilities, who then purchase the products.

Classic sauce One of five sauces also called "mother sauces," since other sauces can be made from them. The classic sauces are béchamel sauce, velouté sauce, espagnole (brown) sauce, hollandaise sauce, and tomato sauce.

Close-out A short-term promotional discount to introduce new products or to quickly sell products with outdated packaging.

Coaching The process of encouraging employees to follow work practices they have been taught.

Comfort food Familiar menu items prepared as a customer might do at home.

Competencies Standards of knowledge, skills, and abilities required for successful job performance.

Competitive bidding The comparison of vendors' prices for products of acceptable quality to determine the least expensive alternative.

Continuous quality improvement (CQI) A system of ongoing efforts to better meet or exceed customers' expectations, and to define ways to perform work with better, less costly, and faster methods.

Contribution margin (CM) The difference between an item's revenue (selling price) and food cost.

Control state (alcoholic beverages) A state that is the sole supplier of spirits. Individuals and retail establishments must purchase all spirits directly from state stores.

Cost of goods sold (COGS) The cost to purchase the products that generate the food or beverage revenue.

Cost-effective Yielding time and money savings that are greater than what an item costs.

CO_2 Carbon dioxide—the colorless, odorless gas used to carbonate water in soft drinks and beer.

Credit memo A document used by accounting staff to adjust information about product quantities and costs included on a delivery invoice to ensure that the operation pays only for the actual products that are acceptable and have been received.

Crew schedule A chart that shows employees what days and hours they are expected to work during a specific time period, usually a week.

Critical control point (CCP) A principle that helps identify and evaluate hazards.

Cross-contact A situation that occurs when one food comes into contact with another and their proteins mix.

Cross-contamination The transfer of microorganisms from one surface or food to another.

Cross-selling Strategies used to promote products and services offered in addition to those noted on a specific menu.

Cyclical (cycle) menu A menu that is planned for a specified time period and then repeated.

Decaffeinated coffee Coffee that has had its naturally occurring caffeine reduced or eliminated entirely.

Decentralized purchasing A system in which department heads may do the purchasing, often with input and approval by the manager, especially for high-cost or large-quantity purchases.

Du jour menu A menu that changes daily.

Edible food yield The usable amount of a food ingredient that can be prepared from a given purchase unit of the ingredient.

Edible portion (EP) The weight of an item after cooking.

Essential function (Americans with Disabilities Act) The key duties that an employee must be able to perform to do the work.

Expediting Interacting with vendors about purchase orders that have been placed but for which products have not been delivered.

Farm-to-fork The flow of food through the stages of growing, harvesting, storage, processing, packaging, and preparation.

Fire suppression system A system containing chemicals that are automatically sprayed on equipment surfaces below the system if a fire begins.

First in, first out (FIFO) An inventory system in which products that have been in storage the longest are the first issued.

Fiscal year Any 365-day accounting period that does not begin on January 1.

Fixed cost A cost, such as rent, that does not change as revenue changes.

Food allergy An allergy that occurs when the body mistakes an ingredient in food, usually a protein, as harmful and creates a defense system (antibodies) to fight it.

Foodborne illness A disease transmitted to people by food. It can be caused by pathogenic microorganisms, chemicals, or physical hazards.

Food cost percentage The percentage of all food revenue that was spent to purchase the food products required to generate that revenue.

Free pour To pour an alcoholic beverage without a portioning tool.

Garnish An edible decoration used to make a menu item attractive.

Guarantee (banquet) An agreement about the number of meals to be provided at a banquet.

Hazard Analysis Critical Control Point (HACCP) system A system used to control risks and hazards throughout the flow of food.

Herbal tea A tea that contains no true tea leaves but is made by steeping the flowers, berries, peels, seeds, leaves, or roots of plants in boiling water.

High check average An operation in which high selling prices offset the labor costs needed to process fresh food items.

Hosted bar An event during which alcoholic beverages are paid for by the event's sponsor.

Income statement A document that summarizes the operation's profitability for an accounting period.

Incremental review Revenue increases beyond current revenues.

Ingredient file A computerized record that contains information about each ingredient purchased, including purchase unit size and cost, issue unit size and cost, and recipe unit size and cost.

Inventory turnover rate The number of times in each accounting period (usually a month) that the quantity of food or beverages in inventory must be purchased to generate the food or beverage revenue for the accounting period.

Investment The amount of money an owner has used to start and operate a business.

Issue requisition A document that authorizes an employee to remove products from storage areas.

Jigger A small shot glass–type tool used to measure the amount of alcohol in drink preparation.

Job description A description that indicates the most important tasks that are part of a job.

Job specification Information in the form of a list of the personal qualifications necessary for an employee to be successful in the position.

Labor cost The money and cost of benefits paid for the work employees do.

Leavening The process in which ingredients produce gases that cause dough to rise.

License state A state that grants licenses to wholesalers, distributors, and sometimes to manufacturers that permit these businesses to sell alcohol within the state.

Line-up meeting A meeting that provides specific information for the shift including special events, new ingredients, or discussion about how to resolve an emerging or recurring problem.

Liquor license A state-authorized permit that allows the license holder to sell alcoholic beverages in compliance with state, local, and federal laws.

Make or buy analysis A procedure to determine whether menu items should be made with raw ingredients or purchased in a convenience form with some labor "built in."

Marinating A tenderizing method that involves immersing food items from one to several hours in liquid made from oil and vinegar or lemon juice seasoned with herbs and spices.

Market form The way a food product is purchased, such as frozen or fresh.

Marking (product) Writing the date of receipt and the purchase unit price on the incoming product.

Master schedule A schedule that shows the number of employees needed in each position and the total number of hours each employee must work.

Measured pour spout (alcoholic beverage bottle) A pour spout that controls the beverage quantity by allowing only a specified amount of alcohol to be dispensed.

Menu classification A similar group of items such as entrées, soups, or salads.

Menu costing The process of determining the food cost to produce all menu items that make up a meal offered at a set selling price when standardized recipes are used.

Menu item file A computerized record containing information about menu items tracked with the operation's point-of-sale (POS) system.

Menu mix The frequency with which a menu item is ordered compared to other menu items.

Menu mix percentage A number obtained by dividing the number of each specific item sold by the total number of items sold.

Menu mix popularity percentage The percentage of total menu items that must be sold for a menu item to be considered popular when sales mix analysis is performed.

Minimum–maximum inventory system An inventory system that indicates the minimum quantity below which inventory levels should not fall and the maximum quantity above which levels should not rise.

Mise en place A French term that means "get everything in place."

Mixed cost A cost that contains both fixed and variable components, such as labor.

Net income The profit made in a business.

Net price The total or per-unit amount paid for products after all discounts have been applied.

Nutrition The science of food and how it affects the health and well-being of the person who consumes it.

Off-premise license A license that allows an establishment to sell alcohol that will be consumed somewhere else.

On-premise license A license that allows an establishment to sell alcohol in the same location where it will be consumed.

On-the-job training Training that occurs when a manager or other trainer teaches job skills and knowledge to one trainee, usually at the work site.

Operating budget A financial plan that estimates revenues and expenses for a specific time period.

Ordering A process that occurs when a buyer makes specific commitments to a vendor for a specific purchase.

Order period The time in days or weeks for which an order is normally placed.

Overpour To use more alcohol than allowed by a recipe.

Overtime The hours for which a wage premium of 1.5 times the hourly rate must be paid.

Package inspection program (employee) A policy that discourages employees from bringing backpacks, shopping bags, and other large packages to work and indicates that packages may be inspected when the employees leave work.

Par inventory system An inventory system that specifies the quantity of products needed to bring the inventory level to an allowable maximum or "par."

Pasteurization (beer) The final step in beer production, which heats the product to 140°F to 150°F (60°C to 66°C) for 20 to 60 minutes to kill any bacteria and remaining live yeast cells.

Performance standard A standard that specifies required quality and quantity outputs and defines the "correct" way to perform a task.

Perpetual inventory system A continuous count of the number of items in inventory.

Petty cash A fund with a limited amount of money that is used to make infrequent and low-cost purchases.

Petty cash voucher A slip signed by the person responsible for the petty cash fund that authorizes the withdrawal of cash from the fund for a purchase.

Physical inventory system An inventory system that involves manually counting the number of each product on hand.

Point-of-sale (POS) system A system that records an operation's sales, product usage, and other important information on a daily, by shift, hourly, or other basis.

Position analysis An analysis used to identify each task an employee must do and explain how it should be done.

Prime costs Food costs and labor costs together.

Production forecast A determination of the quantity of each menu item that will likely be sold during a specific time period, such as lunch or evening meal.

Production sheet (prep sheet) A document that indicates items that will be needed for the shift.

Productivity The quality and quantity of output compared to the amount of input such as labor hours needed to generate it.

Profit percentage The profit or loss expressed as a percentage of total revenue.

Proof A measure of liquor strength that represents the percentage of alcohol in the beverage. Percentage is determined by dividing the liquor's proof by 2 (e.g., a 100-proof whiskey is 50 percent alcohol).

Public bar A bar at which customers can be seated.

Punitive damages A monetary amount that may be assessed if a court finds an establishment's actions showed reckless disregard for a customer's safety. Punitive damages often exceed the amount of compensatory damages.

Purchase order (PO) A document that informs a vendor that the vendor's proposal was accepted and that the order should be delivered.

Purchase requisition A request to purchase products required by a department.

Purchase specification A description of the quality requirements of the products that are purchased.

Purchase unit (PU) The standard size of the package or container in which a product is typically purchased.

Purchasing The series of activities that begins when needs are determined by the menu, and ends after products are served. It includes an emphasis on vendor interactions.

Quality system A system that outlines how quality processes and procedures should be implemented, controlled, and maintained.

Reasonable accommodation A change in the job application process, in the way a job is performed, or to other parts of the job (such as employer-sponsored training and benefits) that enables a person with a disability to have equal employment opportunities.

Receiving report A report used in operations that calculates food costs on a daily basis. The report separates delivery invoice information used for daily food costing.

Recipe conversion factor (RCF) A number used to adjust ingredients in a recipe when the number of servings and/or serving size for a current recipe differs from the number desired.

Recipe evaluation A formal process in which members of the taste-test panel assess whether a recipe produces an acceptable product.

Recipe management software Computer programs that involve or impact standardized recipes.

Reduced oxygen packaged (ROP) Food contained in a package in which oxygen (a) has been removed, (b) has been displaced with another gas or combination of gases, or (c) something else has been done to reduce the oxygen content to a level below that which is normally found in air.

Repeat business Revenue from customers who return because they enjoyed their experience during previous visits.

Request for proposal (RFP) A document that requests prices from vendors for products of a specified quality.

Rollout The process of introducing something new in an operation.

Roux A thickener made from equal parts flour and butter, cooked on the stove at a low temperature.

Safety level The minimum number of purchase units that must always remain in inventory in case of late deliveries or unexpected increases in product usage rates.

Sales forecast report A report that tells the production staff about the overall product needs to meet the sales forecast.

Sales mix analysis A study designed to determine the popularity and profitability of competing items on a menu.

Serving cost The cost to produce one serving of a menu item prepared according to a standardized recipe.

Slack-out seafood Seafood that was frozen and then thawed to appear fresh so it could be sold at a higher price.

Sneeze guard (or food shield) A see-through solid barrier used to protect food in a self-service counter from customers who might cough or sneeze.

Sommelier A service employee with extensive knowledge about wine including wine storage and wine and food affinities.

Specialty line vendor A vendor that provides a narrow product line, but a deep selection within the line.

Spot checking A regular review of standardized recipes and food products as a way to maintain product quality.

Standardized recipe The set of instructions to produce and serve a food or beverage item that will help ensure that quality and quantity standards will be consistently met.

Standardized recipe file A computerized record that contains the recipes for menu items produced along with each item's selling price and food cost percentage.

Stock out A situation in which a product is no longer available in inventory.

Suggestive selling The tactic of using recommendations to ensure that customers know about the products and services offered by the restaurant or foodservice operation.

Supply and demand, law of Economic beliefs about the supply of an item and its price relative to its demand. Generally, as the price of an item increases, the demand for the item decreases (although the reverse can also occur).

Sustainability Activities including water conservation, energy efficiency, and recycling that can lessen an operation's impact on the environment.

Table d'hôte menu A menu that offers an entire meal at a set price.

Table turn The number of times a dining-room table is occupied during a meal period.

Tare allowance A feature on a scale that excludes the weight of a pot or pan placed on the scale to hold ingredients being measured.

Target market A group of people with similar characteristics and similar demands of the marketplace.

Task A duty or activity that is part of a job (position).

Task breakdown A description of how each task in the task list should be done.

Task list A list that specifies all tasks that are part of a job.

TCS food Food that needs time and temperature control for safety.

Temperature danger zone The temperature range at which most microorganisms grow best: 41°F to 135°F (5°C to 57°C).

Tenderizing (meat) Breaking down connective tissue in meat.

Thickener Any ingredient added to a liquid to make it thicker.

Training lesson The information and methods used to present one training session.

Transfer An adjustment to cost of goods sold that increases or decreases food or beverage expense to match product costs with the revenue generated by the product's sale.

Trend A gradual change in customers' food preferences that is likely to continue for a significant time.

Truth-in-menu laws Laws in many locations that require menu descriptions to be honest and accurate.

Underpour To use less alcohol than required by a recipe.

Value The relationship between selling price and quality.

Variable cost A cost that changes in proportion to revenues.

Variance (budget) The difference between an amount of revenue or expense indicated in the budget and the actual amount of the revenue or expense.

Vintage Wines that are grown from grapes in one vineyard during one season.

Well brand (spirits) Spirits that are served when there is no preference for a specific brand.

Wine and food pairing The idea that some wines go better with some food items than others, and that wine should be selected after the food item to match the food.

Wine list A special menu that identifies the wine selections offered along with their selling prices.

Workflow The movement of products through work stations.

Yield The number of servings and serving size specified in a recipe.

INDEX

F

G

H

I

N

Q

NOTES

NOTES

NOTES

NOTES

NOTES

NOTES

NOTES

NOTES

NOTES

NOTES